Construction and Building Automation

This book is intended to be used as a textbook in undergraduate civil engineering and construction courses to introduce cutting edge mechanical, electrical, and computer science topics that are needed for civil and construction engineers to collaborate in inter-disciplinary automation projects.

Part I introduces the basics of hardware and software technologies that are needed for implementing automation in buildings and construction. The content begins with the fundamental concepts and uses practical examples to bring out the benefits of automation through case studies that are easy to understand. No other book uniformly treats the subject of automation within the context of buildings and construction activities. While the technology needed for these two application domains are similar, the unifying principles are not well recognized. This book will bring out the fundamental principles that could form the basis of application to these two domains. For example, it will become clear that sensors, actuators, and controllers, along with smart control strategies could be used for automating tasks within buildings and on construction sites.

Part II of the book will introduce key advances in the areas of machine learning and artificial intelligence that are significant for the intelligent control of buildings and construction equipment. Control algorithms and techniques for data analytics are explained in a form that is appropriate for non-computer science students. Each chapter contains several hands-on exercises meant to apply the principles that are covered. These include numerical problems as well as design and analysis examples.

This new textbook:

- Introduces hardware and software needed for automating engineering tasks
- Presents examples of applications in the control of building systems
- Illustrates of the use of automation for improving construction processes
- Provides a lucid introduction to advanced computing concepts, machine learning, artificial intelligence, and control algorithms to construction and engineering students

It is sure to be essential reading for a growing number of courses in smart construction, building automation, robotics, intelligent buildings, and construction 4.0.

Supplementary material including answers to exercises in the book will be provided on the author's website: https://bennyraphael.com/book2022/

Benny Raphael is a professor in the Department of Civil Engineering at IIT Madras, India.

"A comprehensive treaty on a growing area of knowledge."

Professor Abhijit Mukherjee, *Curtin University, Australia*

Construction and Building Automation

From Concepts to Implementation

Benny Raphael

LONDON AND NEW YORK

Cover image: © Benny Raphael

First published 2023
by Routledge
4 Park Square, Milton Park, Abingdon, Oxon OX14 4RN

and by Routledge
605 Third Avenue, New York, NY 10158

Routledge is an imprint of the Taylor & Francis Group, an informa business

British Library Cataloguing-in-Publication Data
A catalogue record for this book is available from the British Library

Library of Congress Cataloging-in-Publication Data
Names: Raphael, B. (Benny), author.
Title: Construction and building automation : from concepts to implementation / Benny Raphael.
Description: Abingdon, Oxon; New York, NY : Routledge, [2022] | Includes bibliographical references and index. | Summary: "This book is intended to be used as a textbook in undergraduate civil engineering and construction courses to introduce cutting edge mechanical, electrical and computer science topics that are needed for civil and construction engineers to collaborate in inter-disciplinary automation projects" — Provided by publisher.
Identifiers: LCCN 2022002858 (print) | LCCN 2022002859 (ebook) | ISBN 9780367761271 (hbk) | ISBN 9780367761103 (pbk) | ISBN 9781003165620 (ebk)
Subjects: LCSH: Building—Data processing. | Building—Mathematics.
Classification: LCC TH153 .R28 2022 (print) | LCC TH153 (ebook) | DDC 690.0285—dc23/eng/20220413
LC record available at https://lccn.loc.gov/2022002858
LC ebook record available at https://lccn.loc.gov/2022002859

ISBN: 978-0-367-76127-1 (hbk)
ISBN: 978-0-367-76110-3 (pbk)
ISBN: 978-1-003-16562-0 (ebk)

DOI: 10.1201/9781003165620

Typeset in Times New Roman
by Apex CoVantage, LLC

Contents

Foreword vi
Preface viii
Acknowledgments x

PART I **1**

1 Introduction to automation 3

2 Hardware for automation 15

3 Software for automation 51

4 Application: building automation 69

5 Application: construction automation 111

PART II **135**

6 Introduction to machine learning 137

7 Basic mathematics for machine learning 153

8 Regression 179

9 Classification task 198

10 Inductive learning – decision trees and random forests 239

11 Unsupervised learning algorithms 248

Epilogue 264
Index 265

Foreword

In *Building and Construction Automation*, Professor Raphael delivers the fundamentals of this rapidly growing and evolving domain clearly, cohesively and cogently. Senior undergraduates and masters students in civil, architectural, construction, environmental, and mechanical engineering programs will find this text a valuable introduction on which they can build with both experience gained in the field and additional courses in building science, construction robotics, infrastructure computer vision, smart structures, modularization, circular engineering in the built environment, generative building design, and data analytics. Students in mechatronics, and in electrical and computer engineering will use this text to ground and provide a built environment context for their work in intelligent control systems, robotics, machine learning, computer vision, and mixed-reality technologies.

Professor Raphael reduces the overwhelming complexity in this fluid domain without oversimplifying. For example, automation is reduced to hardware and software, and the domain is reduced to the act and the result of building. Civil engineers who have taken a circuits and instrumentation course will know enough to appreciate and absorb the explanation of hardware elements such as power sources, sensors, actuators, controllers, communication devices, and robotics. Having taken an introductory computing course, they will further absorb the importance of wireless and wired communications, the central role of protocols, and the need for control systems, including the PID (proportional integral-derivative) control algorithm that remains ubiquitous in practice, despite the existence of highly sophisticated alternatives.

A humanist theme permeates the text. A smart building is defined as one in which information and communication systems work together with building systems to *maximize the satisfaction of its occupants*. While it is recognized that stakeholders have differing perspectives (to an owner, the building is an investment that must pay off), the role of the occupant is foremost. Construction automation and robotics present similar moral dilemmas. They may improve quality and project performance in terms of cost and schedule; they may even improve safety. Will they take away jobs, or are they an inevitable evolution that will result in more skilled and highly paid jobs? Professor Raphael addresses such concerns head-on in the text, and by doing so keeps us focused on the dignity of the builders and the wellbeing of society.

Humanity recognizes the imperative of sustainability to the world's wellbeing. Buildings require vast amounts of abiotic materials and energy to construct. For example, they demand 50% of global steel production. They are responsible for 40% of the solid waste produced by our societies. Operating them requires up to 40% of the energy produced by our economies. Mass and energy must be manipulated in extremely large quantities in the act and result of building; this is unavoidable. As we seek to reduce this footprint through circular engineering in the built environment, building construction and automation will play a key role. For example, modularization is identified in the text as conducive of automation and as a critical element of design for automation, design for reuse, and design for disassembly. Deep renovations and retrofitting of existing buildings to reduce their operating energy depend on the principles of building automation explained in this text.

In summary, this text supports what should be a required pair of courses in most civil, construction, and architectural engineering programs. It is a substantial contribution to the knowledge base.

Prof. Carl Haas, University of Waterloo, Canada (Former President of the International Association for Automation and Robotics in Construction – IAARC)

Preface

The book is primarily aimed at civil engineers, architects, and project managers. They may not be that much interested in the proofs of theorems and other mathematical details. They want to know what techniques are available and where they can be used. Hence mathematical proofs and derivations are omitted, except for some important concepts in which the derivation or proof is essential to understand the fundamental principles involved. The book is not written for computer scientists or mathematicians. They are likely to be disappointed by this book because it does not have the mathematical rigor that they expect.

However, the book aims to teach fundamental principles rather than software tools. It is not advisable to train engineers to use toolkits as black boxes without understanding the details of the underlying algorithms. They need to be clear about what are the assumptions involved, under what circumstances the techniques can be applied, and the limitations of the techniques. This requires a thorough understanding of the mathematics and the fundamental principles. This is especially true about topics related to machine learning. The primary message from this book is that **not all methods are suitable for all tasks. It requires considerable knowledge and experience to formulate the problem correctly and to apply the right techniques to solve them.**

Since this book is meant to be an undergraduate or postgraduate textbook, exhaustive literature reviews are not provided. The objective is mainly to introduce basic concepts. Occasionally, references are provided to research papers, mostly based on the author's research. These are meant for students who wish to go deeper into application examples that illustrate the concepts described in the book. Special emphasis is placed on solving numerical problems. Working out numerical problems helps in understanding fundamental mathematical concepts. Some of the examples and exercises might be too simple for students who have a strong mathematical background. However, in my experience as a teacher, I find that there is a wide spectrum of students in every class. Some students may not be strong in mathematics, but they have brilliant ideas and make good engineers. Such students need more exercises and worked-out solutions to understand the concepts that seem obvious to more mathematically inclined students. The book is written to cater to the requirements of both categories of students. Students with strong mathematical background can simply skip the examples that are too elementary. Other students get a chance to refresh their knowledge of mathematics without having to search for these topics in mathematics textbooks. Introductory chapters are written without much mathematical content, for example, Chapter 6. These chapters could be used to give an overview of concepts to a non-mathematical audience. The mathematical details are provided in subsequent chapters.

Engineers, architects, and project managers have a considerable amount of specialized domain knowledge that is essential for implementing automation in buildings and construction projects. However, in most colleges, they do not learn automation technologies or machine learning. This type of inter-disciplinary knowledge is essential for working in projects involving automation. They need to collaborate effectively with electrical and mechanical engineers, and computer scientists. Only then we will be able to see appropriate use of automation technologies in building and construction industry.

The topics covered in this book evolved over many years through teaching experience in these subjects at the National University of Singapore (NUS) and IIT Madras. In addition to the formal courses taught at these institutes, the author conducted a number of workshops under the umbrella of the International Association for Automation and Robotics in Construction (IAARC). Many examples and exercises included in this book were part of these workshops, which had significant hands-on content. This book could be used to replicate these workshops in other parts of the world. The numerical problems, review questions, and group discussion exercises could be used to make such workshops lively and interesting. I hope this book will help to spread knowledge of automation to civil engineers and architects, and to enable wider adoption of these technologies.

Acknowledgments

Some of the automation applications described in this book were supported by research grants from various organizations. The author wishes to thank the Department of Science and Technology (DST) and Science and Engineering Research Board (SERB), India, who funded two projects through the grants DST/TSG/AMT/2015/234 and IMP/2018/000224.

Some automation workshops organized by the author were sponsored by the Centre for Continuing Education (CCE) at IIT Madras. Other organizations and colleges who invited the author to conduct such workshops are also gratefully acknowledged. Support from the executive committee of IAARC for various activities of the author helped in preparing this material. Special thanks to Prof. Carl Haas of the University of Waterloo for writing the Foreword.

Many ideas in this book were developed through the research work of my students. Particularly, the students who worked on automation, Ranjith, Aparna, Shanmugaraj, and Sundararaman, contributed to the ideas described in this book. It is gratefully acknowledged that several students and project staff collected images included in this book. Special thanks to my project staff who proofread and corrected mistakes in the manuscript.

Thanks to everyone who supported me in my professional activities, the Department of Civil Engineering at IIT Madras, and especially my colleagues in the Building Technology and Construction Management (BTCM) division. The Institute of Eminence Research Initiative Project on Technologies for Low-Carbon Lean-Construction (TLC2) that we have recently started has a strong emphasis on construction automation and has provided financial support for research in this area.

Part I

1 Introduction to automation

1.1 Introduction

A fictional scenario

Ajay has a job interview at Intelligent Building Consultants, Chennai. This is his dream job, and he was always looking forward to this during his four years of undergraduate education at IIT Madras. This company is famous for implementing many smart buildings in India. He is excited about working there. He reached the headquarters well ahead of the scheduled time of the interview. At the entrance, the doors opened automatically, which he thought that was very convenient, especially since he is carrying lot of documents, and his hands are not free. Right at the entrance, there is a kiosk with an electronic display showing the map of the building and the directory. He could search for the location where he was asked to report. Next to the electronic display, there is a place to keep his documents so that he could operate the kiosk conveniently. It looked like the designers of the building had carefully thought about all the little details needed to make the building comfortable to all the users. He typed in the words HR, and immediately the location was shown on the map, and the way to the HR department was highlighted. He moved towards the lift and as soon as he started walking towards it, the lift started coming down. There is a camera detecting that he was moving towards the lift and responded promptly. Inside the lift, there is a voice recognition system. It prompted him with a voice message, "Where do you want to go?" Ajay calmly replied, "HR department". The system understood his speech, which is not surprising today. But it was amazing that the lift had knowledge about the building! It knew on which floor the HR department was located. It asked for confirmation: "You want to go to HR department on fifth floor?" Ajay agreed and the lift started moving. As the lift door opened, he noticed that the lights turned on automatically. It had been switched off earlier since there was no one in the lobby. Ajay had a quick look through the window of the lobby. It was a scenic view with lot of greenery outside. The window blinds were closed on the west side because there is direct solar radiation coming from this side, but there is a button to manually open the blinds if required.

The main door at the entrance of the HR department opened automatically because an image recognition system identified the candidate. The system was expecting him because his interview schedule had been input into the database. The HR manager is informed automatically that the candidate has arrived. A voice message greeted him and asked him to wait for a while. Finding the path had been quite easy and he had arrived ahead of the schedule. The TV is turned on for him at the reception. Lights and air conditioning are turned on to comfortable settings. An LCD display shows the energy consumption of the building. He is impressed by how much the company cared for the environment.

DOI: 10.1201/9781003165620-2

The HR manager took him to the meeting room which had been kept ready for the interview. The room had been booked through a facilities management system that was smart enough to turn on the air-conditioning ahead of time to keep the room cool. Ajay joins the video conference with the CEO of the company, who is based in a foreign country. The room has a large monitor, and it was almost like speaking to him face-to-face. He is impressed by the clarity of video and audio. The interview went smoothly. The company is keen to take this brilliant engineer from the most prestigious engineering institute in India. The young engineer is more than happy to accept the offer. When he leaves the HR department, he has a new reason to join the company – just for the pleasure of working in that building.

This is not a real story but created to illustrate how automation can enhance comfort and convenience. The present generation is exposed to the conveniences of automation, whether it is for withdrawing money from the bank or finding the route through voice commands on a mobile phone. They expect similar facilities from buildings, but many buildings have not kept pace with the requirements of this generation. Even building construction techniques have not kept pace with the developments in automation and artificial intelligence. Many buildings are still constructed with brick and mortar using age-old construction methods, mostly using manual labor.

Can we change this scenario? This book discusses how this can be made possible. It introduces both hardware and software technologies that are needed for achieving this. It is expected that with the spread of this knowledge, there will be better adoption of automation both at the construction stage and at the operation stage. This book is intended to be used as a textbook in undergraduate civil engineering and construction courses to introduce cutting edge mechanical, electrical, and computer science topics that are needed for civil and construction engineers to collaborate in inter-disciplinary automation projects.

This book uniformly treats the subject of automation within the context of buildings and construction activities. While the technology needed for these two application domains are similar, the unifying principles are not well recognized. This book brings out the fundamental principles that could form the basis of application to these two domains. For example, it will become clear that sensors, actuators, and controllers, along with smart control strategies could be used for automating tasks within buildings and on construction sites.

1.2 Building automation

The main goal of building automation is to make buildings smart. A smart building behaves intelligently, just like a smart robot exhibits intelligent behavior. Today, it is not difficult to imagine a building that has artificial intelligence (AI); AI has percolated into every aspect of our lives. However, it is necessary to review some basic concepts and definitions related to these.

1.2.1 Intelligence

When do we say that something or someone is behaving intelligently? If someone claims that he can read and write, add, and multiply numbers, no one will consider that as impressive. These are basic skills that are expected from every adult. But if a two-year-old does complex math, we will undoubtedly consider that as intelligent. The notion of intelligence is relative. We judge intelligence from achievements that are not normally expected from people or things. What is considered as smart today may not be considered so tomorrow, when smart technologies become more widespread. Nevertheless, intelligence will always be detected through performance comparison with what is commonly found in the society.

This discussion is true about machine intelligence as well. Autonomous vehicles captured the imagination of people in the 1990s. However, they are common today, and people already consider that as mature technology. Common people do not consider self-driving cars as intelligent; their expectations have grown much higher. This poses a problem; how do we define intelligence? How do we categorize something as smart? We need a working definition so that clients and designers of buildings do not have a mismatch in expectations.

We first look at some concepts in computer science. Artificial intelligence is an area of computer science that attempts to mimic human intelligence in computer systems. There have been many heated discussions about what true intelligence is; this issue is still being debated. However, certain concepts have emerged. First, systems should learn from experience – knowledge should not be static. The system should have mechanisms built into it that allow accumulation of knowledge throughout its life. Second, the performance of the system should improve with time. The system should be able to adapt its behavior using new knowledge accumulated through experience. Third, the system should exhibit autonomous behavior. It should be able to make decisions by itself instead of relying on human beings. All three aspects are true about human intelligence as well. We learn continuously by acquiring knowledge from other people, books, and by self-discovery. We adapt our behavior based on experience; we try not to repeat the same mistakes. Smart workers do not need to be told every step for carrying out a task, they work autonomously with minimum instructions from superiors.

Many examples of systems that have evolved from research in artificial intelligence are found in the world today. Intelligent robots have been demonstrated to perform tasks which were considered to be impossible a couple of decades ago. Autonomous cars have already been tested for thousands of kilometers and people are ready to accept them. These technologies are likely to cause major disruptions in the society. It could change the way buildings and cities are designed. Do we need hundreds of parking spaces in a commercial building? Can a self-driving car drop people at the building, park itself at a convenient location nearby, then come back and pick up the owners when they want to return? If autonomous vehicles can improve traffic flow, reduce the amount of energy required, and save time for people, should not the roads be designed to facilitate this?

1.2.2 Smart buildings

The concept of intelligent buildings originated in the 1980s. It denoted buildings with sophisticated telecommunications, building management, and networked services. The viewpoint of this era was that a high-tech building is an intelligent building, one with a computerized operating and management system as well as the latest hardware. Later, people argued that buildings can never be intelligent. High technology will never make a building intelligent. The term *smart building* started getting wider acceptance. There have been many attempts to define what is smart. However, changing the term from *intelligent* to *smart* does not make any system any different. Therefore, in this book, the two terms will be used interchangeably.

Exercise 1.1 Group discussion

Discuss whether your home is a smart building. Write down five features that you expect to see in a smart building. Discuss if you have come across any commercial building in your city which has these features. How widespread are smart buildings today?

Building owners expect several features in a smart building. They expect the capability to monitor and control building services such as HVAC, lighting, lifts, fire alarms, access, and security. Environmentally conscious clients expect reduction in energy use of buildings during the operation phase. However, they do not want this at the expense of comfort. They want to maximize comfort conditions, maintain steady temperature set by the user, control relative humidity, ensure adequate fresh air supply, etc.

Another recent trend is the use of automated appliances and devices inside buildings. These perform tasks without manual intervention. While automated light switches and automatic doors have been there for a long time, more recent developments have been to control the operations of home appliances such as refrigerators, washing machines, and cookers. Won't it be convenient to find out what is inside your fridge when you are in the supermarket so that you can buy things that are low in stock? Is it not convenient if the dishwasher keeps us posted on the status of the cycle and informs when detergent starts running low? While you are waiting in a traffic jam, do you want to turn on your cooker to keep your food warm and ready when you arrive home (assuming of course, the food is already inside the cooker)? To extrapolate this a little bit, do you want a robot in your kitchen to prepare the food for you while you are returning from work?

The ability to control many of these devices with smart phones is already there in many commercial products. Security devices such as cameras are being widely controlled through mobile phones. While many users are not concerned about the energy savings through smart building systems, there is no one who is not worried about safety and security. It seems that this market is likely to grow rapidly since it is closely connected with the fundamental human concern about safety. Even if it costs ten times the amount needed to implement daylight harvesting that helps to reduce energy required for electrical lights, people are willing to install security systems to monitor the opening of every door and window from their place of work.

Artificial intelligence has improved the capabilities of many of these devices. Face recognition has matured enough to be used for access control. You no longer have to carry your keys with you to enter your building. The building can recognize your face, fingerprints, voice, and gestures. Even if one technique does not work, alternative methods could be used for confirmation of the identity. The doors can be opened by simply shouting at a smart voice assistant that can recognize your voice. Big IT companies have invested heavily in home automation, and this sector is likely to witness huge leaps in capabilities soon. Now, with openly available interfaces, device makers will be able to build voice assistants into headphones, smartwatches, fitness trackers, and more.

1.2.3 Role of software

The previous section introduced the idea that automation has the potential to make a building smart, in the sense that there is improved performance. The occupant comfort is enhanced, energy consumption is reduced, safety is improved, and the systems are most usable and convenient. The result of all this is that productivity of the occupants can be potentially increased. However, there is a limit to the improvement in the performance by introducing automation in individual subsystems. Integration is essential for taking performance to the next level. There is better efficiency if all the services can be operated from a single point of control. This also helps to provide coordinated response from various systems to problems as they arise. For example, when there is a fire, all the subsystems have to respond in a coordinated manner; the air conditioning system should close the dampers to prevent smoke from getting transported from one zone to the other; the fire alarms should sound, exit lights blink, and the lifts are disabled (if they are designed for use during a fire).

Integrated coordinated action of different subsystems is possible only using carefully designed software. This software is called a building management system (BMS). In the early days, BMS dealt with only the HVAC system. Other systems were not integrated with the HVAC systems, mainly because there are specialist subcontractors for each subsystem, and they are reluctant to communicate with each other. They think that integration might introduce problems into their systems that are difficult to diagnose and isolate. This line of thinking resulted in separate software systems for managing air conditioning, lighting, fire safety, etc. Even the communication networks were duplicated because subcontractors did not want to expose their networks to others. Now, owners and facilities managers are increasingly convinced about the need for integration, and they are pushing for smart systems that can interoperate seamlessly and function in a coordinated manner. With this objective, BMS should integrate all the building systems, and all subsystems should talk to each other, using a common language.

The term Building Automation System (BAS) is being commonly used almost synonymously with BMS. The term is used mostly by hardware manufacturers and system integrators. The term *building automation* evolved from industrial automation in factories, where the operation of large industrial plants is automated using sensors and actuators. Since most of the hardware required for automation are the same whether it is used in factories or buildings, companies that originally developed the hardware for factories entered the building automation market. Consequently, many terms that were used in industrial engineering came to be adopted in building systems also.

It is time to clearly separate the functions of the BAS and the BMS. In this book, BAS refers to low-level software that is used to control hardware devices, that is, focusing on implementing automation in buildings. The term *low-level software* is more commonly used in this book instead of BAS. BMS is the piece of software supporting managers, rather than system integrators and technicians who implement automation. BMS is a higher-level software with management focus compared to BAS. With this separation of the roles of the two systems, much confusion can be avoided.

Another development is the emergence of facilities management as a separate industry. Large facilities such as airports and commercial malls involve complex operations and maintenance procedures. It is impossible to manage all these without sophisticated software. The software required for this evolved independently of automation systems and was given a different name, Facilities Management System (FMS). Initially, the focus was on efficient utilization of space and inventory. Soon it was realized that the management of the facilities cannot be separated from the building systems. Then, developers started introducing and duplicating some of the features of BMS into FMS. Eventually, BMS and BAS will become subsystems of the FMS as shown in Figure 1.1.

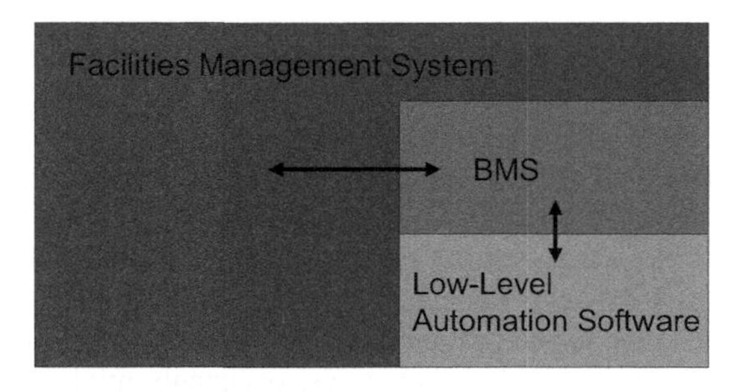

Figure 1.1 Software for building management

In Figure 1.1, BAS is the bottom most layer, labeled as "Low-Level Automation Software". It refers to systems for implementing lighting, fire alarms, access control, etc. This is a device-centric software supplied by hardware vendors, meant for technicians and software developers to implement automation.

At the next level is the BMS, a software that coordinates the operations of different building systems. Typically, it has a central database and a communications network to integrate electronic systems that control building services. The database stores settings such as set points for environmental parameters, historic data related to operational parameters, occupancy schedule, etc. The building manager is able to set the parameters and analyze the performance of various building systems.

The FMS is at the top level. It has many functions related to the management of the facility. It records location and details of equipment and assets, keeps track of their maintenance and related issues, analyzes space utilization, etc. It also provides an interface to the BMS so that managers are able to perform tasks without knowing too much technical details. It is a tool for managers rather than for technicians.

The trend behind integrating FMS with BMS is the realization that building operations should be seamlessly integrated into the workflow of an organization for better effectiveness. Workflow refers to the "flow" of information through various stages in performing a task (which is defined within a business process). For example, an organization might have policies and procedures for booking a meeting room, which are typically defined as a workflow within a management software. When a meeting room is reserved for a function, we also need to ensure that air conditioning, lighting, and audiovisual systems are performing well. It is not nice to find out that the meeting room is not in good working condition after it is reserved, and all the arrangements have been made for the meeting. If the FMS is linked to the BMS and the BAS, there is seamless exchange of information between the different systems, and smart performance can be achieved.

To conclude, building automation is heavily dependent on different types of software in addition to the hardware. The "smartness" of the building depends to a large extend on the level of sophistication of this software.

1.2.4 Smart building: a formal definition

In the previous sections, it was argued that the smartness of a system is judged by its performance. The performance of any system can be enhanced using both hardware and software, with appropriate levels of automation and integration of subsystems. Since buildings are created and exist for their occupants, the most important criterion is how well they meet the needs of the users. Based on this, a working definition is given here:

> A smart building is one in which information and communication systems work together with building systems to maximize the satisfaction of its occupants.

This is an adapted version of the definition of intelligent buildings given in some books. The definition just stated is user-centric, that is, the emphasis is on the needs of the occupants. The needs change with the value systems of the occupants. Some owners consider environmental sustainability as a crucial parameter. In that case, the building systems and software should aim to reduce energy consumption and optimize other parameters that are important for achieving sustainability goals. In other cases, occupants might consider comfort and convenience as more important than sustainability since they affect their productivity and efficiency. In this case, the buildings systems might be geared towards this goal. Many other performance parameters might be found to be relevant: use of resources, embodied energy, carbon footprint, life-cycle costs, etc. All these could be considered as parameters that affect the satisfaction of the occupants. In any case, the integration of information and communication systems with physical hardware is an essential part of the definition because this is what historically differentiated an intelligent building from an ordinary building.

1.2.5 Beyond buildings: smart cities

Even though, the focus of this book is primarily on buildings and construction, it is useful to think beyond buildings to view a broader picture. Society is increasingly concerned with issues such as sustainability and climate change. Building performance cannot be isolated from these issues. We should also think about the performance of the city (or a wider region) as a whole. The concept of a smart city emerged with this line of thinking. A smart city uses information and communication technologies (ICT) to enhance the performance of urban systems, including, infrastructure, finance, and governance. Just like a smart building, it focuses on improving comfort and convenience of all the occupants. This is attained by both software and physical infrastructure. Needless to say, both software and hardware automation contribute to making a city smart.

1.3 Construction automation

The goal of construction automation is to make the construction processes smart. Therefore, both building and construction automation share similar goals that are linked to system performance. Smart construction processes reduce the time and cost of construction, eliminate material and other types of waste, and enhance the satisfaction level of all the stakeholders. Stakeholders have different priorities. Owners are interested in a quality product. Project managers focus on timely completion of the project within budget. Workers (and managers) are concerned about safe working conditions. Architects and designers want to ensure that the finished product meets their specifications and expectations. Automation must satisfy all these needs.

It is now generally accepted that construction automation will not only help in decreasing the time of construction and amount of labor required but also reduce safety hazards and the risks involved. Many manual processes have already been automated, especially for heavy infrastructure projects. Prefabrication has also increased in many countries, and many construction activities have been shifted to factories from construction sites. Advanced robotic systems will help fabricate building elements in factories just as machine components and automobiles are manufactured. Such large-scale automation might be necessary to keep up with the demand for housing and infrastructure in developing countries. Also, this has the potential to make housing affordable because different types of wastes in construction significantly increase the costs. For example, expensive equipment might lie idle because of poor planning and coordination; workers might be wasting their time because material and tools do not arrive on time. Shifting activities to controlled factory-type environment helps to eliminate many such wastes.

1.3.1 What can be automated?

Construction involves tasks that are executed physically (such as placing bricks and mortar), as well as processing of information (for example, transmitting the status of activities from sites to offices). Construction projects might be delayed because of inefficiencies in both types of tasks. Therefore, automation of physical tasks as well as information processing require equal attention. Traditionally, it has been easier to automate information processing since information technology has undergone unimaginable progress during the last few decades. Progress in automation of physical tasks in construction has been slow, but there is renewed attention to this, especially because of the demonstrated successes in manufacturing. Both types of automation should be adopted at appropriate levels to produce overall improvements in productivity and efficiency in construction. This book covers issues related to both the type of tasks.

In industrial engineering, automation refers to the use of programmable devices for performing tasks without manual intervention, that is, using computer control for mechanical tasks. Strictly applying this definition, many of the technologies adopted in construction qualify only as mechanization and cannot be considered as automation. For example, the use of motorized hoists for lifting construction material is only mechanization, not automation. However, tools and equipment are getting more sophisticated, and many of them have embedded systems with microprocessors. Even a cutting machine might have a controller that uses feedback for the safe operation of the device. Hence, the distinction between mechanization and automation is getting blurred. Even if end users are not able to program the equipment, they might have inbuilt programs for the correct sequence of operations, that is, operations that are automated with computer control.

1.3.2 Automation of physical activities

Researchers have attempted to classify tasks into standard types depending on the kind of movements that are required. Everett and Slocum (1994) argued that all construction operations could be categorized into ten basic tasks. These are: connect, cover, dig, finish, inspect, measure, place, position, spray, and spread. This list can be expanded to include missing basic tasks such as hammering, cutting, bending, etc. Potentially, all these basic tasks can be automated. Machines can also be programmed to perform high-level tasks that involve combinations of these basic tasks. However, it is difficult to develop cost-effective general-purpose machines for any of these tasks because the specifications of the machines will depend on the detailed characteristics of the tasks. For example, connecting two elements must consider the type of forces that are transmitted through the joints. It would depend on factors such as the size and type of objects that are connected, their geometric and material properties, etc. Even though, general purpose industrial robots have been developed that can perform multiple tasks, we are far away from having machines that can perform all the tasks in construction. Therefore, we have to design specialist machines for each category of tasks. This requires studying the kinematics (movements) as well as dynamics (forces) involved in each task.

Exercise 1.2 Group discussion

Watch videos of popular humanoid and - robots that are available on the internet. Discuss what activities might be performed by these robots on construction sites. What are the advantages of using such robots in construction? What are the barriers to their adoption?

Exercise 1.3 Group discussion

Discuss what types of machines are needed for performing the following activities efficiently on a construction site:

1 Transporting materials vertically
2 Transporting materials horizontally
3 Placing material at precise locations
4 Cutting and removing material
5 Alignment and adjustments of building elements
6 Connecting steel parts
7 Painting
8 Plastering

Discuss what are the difficulties in automating these tasks.

Requirements for automation can be simplified by shifting most of the construction activities to factories, that is, through prefabrication. Prefabrication can be done in on-site factories or off site. In the controlled environment of a factory, repetitive movements needed for fabricating large number of components can be achieved through specially designed machines or general-purpose robots. This is already common in automobile factories and other manufacturing industries. Already, automatic machines are available for manufacturing concrete blocks in large quantities at a fast pace. The next level of automation will be manufacturing larger and more complex building elements. These elements must be carefully designed such that assembly of these elements on site is also efficient. Modular construction is an extreme form of this idea. Large modules such as whole rooms are prefabricated in factories, transported, and assembled on site. Such technologies will shift the labor requirements from unskilled workers to skilled technicians. More labor-intensive jobs are done by machines; work that require intelligence and skill are left to technicians. From social sustainability, this is attractive – keeping uneducated workers doing menial jobs is not good for the development of the society.

Even if most components are fabricated in factories, site work cannot be eliminated completely. Many activities must take place on site. Surveying used to be done mostly manually. Now drones are available for this task. Drones equipped with laser scanners can produce a topographic map quickly. Excavators with increasing level of automation are becoming available. These might take a model of the terrain to be excavated, and the paths and activities needed could be determined automatically. Lifting and accurately placing the components on site could be done by autonomous cranes. Warehouse industry might provide many solutions that are needed on construction sites. There are warehouses that are completely automated; objects are picked, tagged, and stacked such that they can be retrieved easily and automatically. Tracking technologies help monitor the movement of materials on site. This data could be transmitted to central servers to help in logistics management.

1.3.3 Automation of information processing

Information technology (IT) is all about how we represent information and how we process and transmit information. However, we have not been able to precisely define what is *information.* For practical purposes we can consider information as consisting of both data and knowledge. Knowledge is considered as relationships in data (Raphael and Smith, 2013). Extracting knowledge from data is the subject of Part II of this book, machine learning.

The pre-construction phase of projects involve quite a bit of information processing. This includes designing and planning, which are information intensive tasks. Such tasks are widely covered in books on computing and software engineering. Therefore, these are not given much emphasis in this book. We focus mostly on the construction phase of projects.

A lot of data are generated on construction sites. This includes manual observations as well as data from sensors. Construction managers need information about site conditions, progress of work, deviations from contractual conditions, etc. This data needs to be represented and stored in a form that computers can efficiently process for effective project management. Automation of information processing involves collecting, representing, storing, transmitting, and analyzing this data. For this, we need to identify what sensors could be used, where they should be placed, and what algorithms are most appropriate for analyzing the data. Sensors that are needed for automation are discussed in Chapters 2 and 5. Algorithms are discussed in Chapter 3.

Digital fabrication is a concept that is gaining lot of attention. In the past, site engineers used to rely on two-dimensional paper drawings. Three-dimensional models help in visualizing the construction better. Issues are detected faster than using paper drawings. In fact, detection of conflicts is now a standard feature of many modeling software. In addition to geometry, other information about the constructed objects is also now stored in digital models using the concept known as building information modeling (BIM). An accurate 3D representation of the objects along with information such as material properties, quantities, costs, and time is possible using BIM technology. This model could be used on site using portable devices to ensure that work is carried out correctly without errors. The next step of automation will be to use these digital models to perform automated activities. Updating of the status of the activities, generating look-ahead schedules, etc. can be done using the digital models along with site data.

1.4 Implementing automation

It was explained that both hardware and software have equal part to play in implementing automation. What hardware is needed? This is discussed in more detail in Chapters 2 to 5. Software needed for automation includes conventional procedural programs for automating computing tasks as well as advanced systems that make use of AI and machine learning. Some conventional programs for building management are discussed in Chapter 4. The role of machine learning is brought out in Chapters 6 to 11. A quick introduction to hardware and software is provided in this section.

1.4.1 Hardware

An automation system is like a human body. The human body has sense organs: eyes, nose, ears, etc. These send signals to the brain. The brain processes these signals, performs analysis, and instructs body parts – arms, limbs, and other organs – to take necessary actions. In an automation system, the controller takes the role of the brain. It is a piece of hardware that is just like a small computer. It has a microchip inside to perform computations. In addition, it has input and output channels to which sensors and actuators could be connected. Sensors read environmental and other parameters. They transmit this information to the controller through the input channels. Actuators take actions such as turning on a device or applying a force. The controller sends commands in the form of electrical signals to its output channels; the actuators that are connected to these channels interpret the signals and perform the actions.

In addition to controllers, sensors, and actuators, an automation system relies on the communication network. At the lowest level, the sensors and actuators need to communicate with the controller. At a higher level, controllers have to communicate with each other for coordinated action. Different network architectures have been developed for efficient and cost-effective communication. Part of the difficulty in implementing automation is because device manufacturers adopt different communication schemes that are often closed and proprietary. The control system should also communicate with humans. Humans have to provide input and they need to see the output. Since humans typically interact with the automation system through a computer, controllers must have networking capabilities to communicate with computers using standard protocols.

Whether the automation system is for controlling building systems or construction equipment, the hardware required are similar. The sensors used might be different, but they communicate using similar interfaces. The actuators might also be different, but the underlying engineering principles are the same. Therefore, both building automation and construction automation can be treated in a uniform manner.

1.5 Implementation issues

While automation brings lot of benefits to users, we are not seeing significant adoption of automation because of many reasons. Some reasons are technical, while others are related to human behavior and psychology.

Historically, there has been strong resistance to mechanization and technology adoption. It was always argued that automation will result in job losses. In many countries, labor unions have protested even the introduction of computers in offices. It took a long time to get computers adopted in government offices and banks. Today, it is accepted that computers are essential for office work and financial transactions; it is impossible to meet the demand without computers. However, people still object to the use of machines in factories and on construction sites. Arguments that robots will replace workers invoke strong sentiments. Benefits of automation get drowned in the arguments related to job security of workers.

Another barrier to the adoption of automation is that capabilities of automated systems do not match users' requirements. In the Charlie Chaplin movie, *Modern Times*, a vendor wants to sell a machine to feed the workers, arguing that it reduces the time needed for lunch break. Definitely, this is not something workers want. In the early days of building automation many products were developed that did not match the expectations and needs of users. Automated light control systems did not consider the fact that there are blind spots where sensors could not detect occupants, and sometimes occupants remain still. Automated blinds irritated users by constantly moving up and down in response to the amount of light coming from the sun. Sufficient efforts were not put into accurately gauging user acceptance and preferences before introducing such products. Many products failed because the first negative impressions created biases among the entire population of potential users.

There are technical barriers also. Electrical components are more likely to break down compared to building elements such as beams and slabs, especially when used outdoors. The life span of a building is more than fifty years. Electrical systems have a much shorter life span. Control systems that are exposed to the environment are more likely to fail soon. Therefore, many architects prefer not to use building elements that are actively controlled, for example, shades that are actuated using electrical systems.

Automation engineers have always struggled with the problem of interoperability. Integration of equipment and computers is not always easy. Standards are only emerging, and vendors do not always provide standard interfaces. For a long time, companies provided proprietary interfaces and made it difficult for other companies to interface with their systems. Open standards have now been developed for most domains, but the problems have not disappeared altogether.

The issue of interconnectivity of sensors and actuators has been given serious thought only recently. How do we enable large number of sensors and actuators to seamlessly connect and exchange information without complex procedures for configuration? The concept that is gaining prominence is the Internet of Things (IoT). The IoT is a large network consisting of devices and people, a platform which makes it possible to access data from anywhere and everywhere. Any device can potentially be connected to the internet and exchange data with others. It has become an important buzzword and many companies are competing to establish dominance in this emerging domain. It is expected that billions of devices will be interconnected through IoT and will enable the availability of large amount of data. This data can be used for better management of resources and would result in high performance smart systems.

1.6 Summary

The need for automation in buildings and on construction sites is brought out in this chapter. Automation is needed for introducing smart behavior into building systems and construction processes. The terms smart systems and intelligence are defined. The questions – What can be automated? and How automation can be implemented? – are briefly discussed.

References

Everett, J. G. and Slocum, A. H. (1994). Automation and Robotics Opportunities: Construction Versus Manufacturing, *Journal of Construction, Management Engineering and Management*, 120, pp. 443–452.

Raphael, B. and Smith, I.F.C. (2013). *Engineering Informatics: Fundamentals of Computer Aided Engineering*, second edition, Chichester, UK: John Wiley.

2 Hardware for automation

2.1 Introduction

Hardware requirements for automation were briefly discussed in Chapter 1. In Chapter 2, automation hardware is discussed in more detail. Several electrical and mechanical components are essential for implementing automation. They appear in the bill of quantities and influence the project cost. Electrical and mechanical systems form a major part of the cost of many building construction projects, especially commercial and office buildings. Hence, project managers and architects working on construction projects that involve automation should be familiar with them. Project managers who coordinate the activities of consultants should be aware of the issues and challenges they face in the projects. Knowledge of electrical and mechanical systems help them in project coordination. Students and researchers who develop prototypes of automation systems need to have deeper understanding of the hardware. Today, many students undertake such projects either as a hobby or for demonstration of new ideas. Demonstration projects are immensely useful for convincing professional companies about the technical viability of new concepts and encouraging them to invest in developing commercial products. This chapter caters to both groups of people: professional architects and project managers, as well as students and researchers who undertake demonstration projects.

The following elements are covered in this chapter:

- Sensors
- Actuators
- Controllers
- Communication networks

This is a multi-disciplinary topic involving civil, mechanical, and electrical engineering. Basic concepts related to hardware can be appreciated even without an in-depth knowledge of these subjects. However, fundamentals of electrical circuits are necessary for understanding how automation systems operate and how they are implemented. Hence these are briefly reviewed first.

2.2 Fundamental of electrical circuits

Electrical currents are created by the movement of charged particles; these charged particles are electrons in commonly used electrical circuits. Electrons move from a point in the circuit to another if there is a potential difference between these points. The potential energy of the

DOI: 10.1201/9781003165620-3

electron helps to overcome the energy required to overcome the resistance in the movement. This is similar to an object moving from a higher position to a lower position because there is a difference in the gravitational potential between these two positions. *Volt* is the unit used to measure the potential difference between two points in a circuit. *Ampere* is the unit of current, which is the amount of charge that moves through the cross section of the conductor in unit time.

An electrical power source such as a battery or a generator produces a potential difference that can be used to drive a current between two points. Chemical batteries convert chemical energy into electrical energy through the reactions that take place inside the cell. Since electrons are released on one side of the cell where the reaction takes place (cathode), there is a negative potential on that side. The other side is at a positive potential (anode). When the two sides are connected by a conductor, electrons move from the cathode to the anode and the resistance to its movement in the wire is overcome by the potential difference. We represent such an arrangement using a schematic such as in Figure 2.1. V represents the potential difference of the power source, R the resistance of the circuit, and S is a switch which can be used to open or close the circuit.

A voltmeter measures the electric potential (voltage) between two points in a circuit. An ammeter measures the current (in amperes or milliamperes). A multimeter is a common instrument used by electricians to measure current, voltage, resistance etc. (Figure 2.2). It has two leads pressed against two points in a circuit to measure the voltage or resistance between the points.

If V is the potential difference between two points and R the resistance of the circuit between the two points, then the current I flowing through the circuit is given by:

$$V = I R \qquad \textit{Eq 2.1}$$

The higher the voltage, the higher the current. The higher the resistance, the lower the current. Knowing any two parameters, we can calculate the third one, provided that the values are steady (do not change with time). There is another way to look at this equation. Pushing a current I through a resistance R causes a drop in potential V across the ends of the conductor.

Polarity refers to the relative direction of positive and negative ends of a circuit. In a chemical cell, the polarity is constant, that is, a terminal is either positive or negative. If a conductor is used to connect the two terminals, the direction of current in it is the always same; the battery produces a direct current (DC). On the other hand, when power is produced

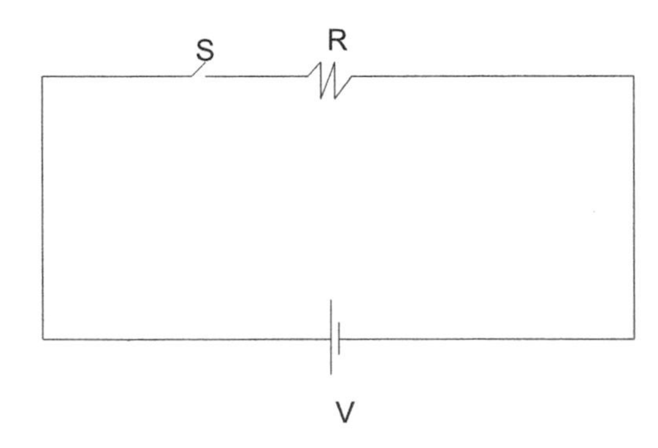

Figure 2.1 An electrical circuit

Figure 2.2 A multimeter

by a generator by rotating a coil in a magnetic field, the potential difference between the two ends changes with time. Typically, it increases and decreases according to a sinusoidal curve. The polarity reverses at regular time intervals, and the current reverses direction periodically (typically 50 times per second, that is with a frequency of 50 Hz). This is called an alternating current (AC). Alternating current is normally used in the electricity grid of most countries for power transmission to homes and factories. This is due to two reasons. First, all the big generators and turbines use the principle of electromagnetic induction for generating alternating current. Second, power transmission losses are lower because voltage can easily be stepped up using transformers.

The power consumed by a part of a circuit is calculated using the formula:

$$P = V\,I \qquad \textit{Eq 2.2}$$

If V is measured in volts and current in amperes, the product gives the power in Watts. This is the energy (per unit time); therefore, the energy is calculated by multiplying power by time. Joule or kWh (kilowatt-hour) are units used to express energy. Energy is consumed in a circuit to overcome the resistance to the movement of the electrons through the wire. This energy usually gets converted into heat and is dissipated by the circuit. Using Eq 2.2, power can be written in terms of either voltage or current as follows:

$$P = I^2\,R \qquad \textit{Eq 2.3}$$

The power consumed by the circuit is proportional to the square of the current. This principle is used to reduce transmission losses when power is transmitted over large distances. From Eq 2.2, if the same amount of power is transmitted, the current I_2 drops if the voltage is increased from V_1 to V_2 according to the equation:

$$P = V_1\,I_1 = V_2\,I_2$$

Therefore, from Eq 2.3, the power dissipated in the circuit is reduced because of lower current. It can also be seen from the equation that the potential drop across the wire transmitting the power reduces according to I^2 R. That means, more power is consumed by the actual load, and less power is dissipated and lost by the wires. This is shown in Figure 2.3.

In alternating current circuits, the voltage and current varies with time, and there might be a phase difference in the sine curves representing voltage and current. That is, the voltage and current may not reach the peak amplitudes at the same time. This timing difference is called the phase shift and is measured in angular degrees. To calculate the average power consumed by the circuit, the instantaneous voltage and current should be multiplied and averaged over time. Therefore, the product of the peak voltage and current will not give the power. The average power P_{avg} is calculated as

$$P_{avg} = V\, I\, \omega \qquad \textit{Eq 2.4}$$

where ω is the power factor (a number between 0 and 1) depending on the phase shift between the current and voltage.

Even though the power supplied by the electrical grid is AC, most electrical devices internally use only DC. Devices such as computers and office equipment have power supply modules inside them that convert AC to DC. A certain percentage of power is lost during this conversion. Today, many buildings have photovoltaic cells installed on their roofs and generate their own power from the sun. The solar cells produce DC power. Since solar power is not available all the time, power supply in buildings use the electricity grid as a backup. Therefore, the DC power from solar cells is converted into AC and fed into the electric circuits of the building. The conversion from DC to AC involves losses. This is compounded by the losses in converting AC back to DC for use inside electronic equipment. To avoid these conversion losses, some modern buildings use only DC power. All the electrical equipment

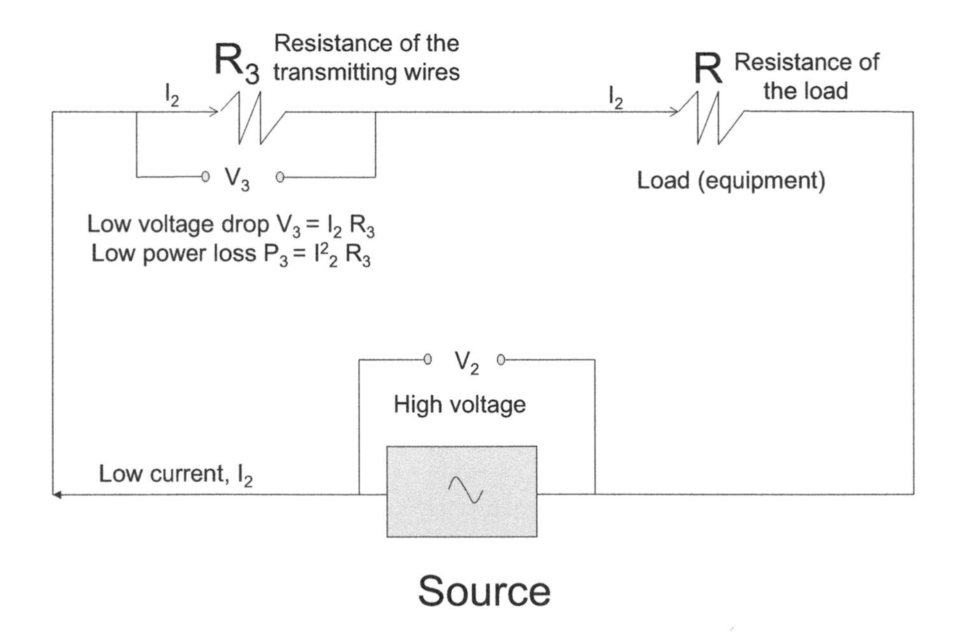

Figure 2.3 Transmission power loss

is designed to work with a DC power supply. Such techniques are helpful in improving energy efficiency and implementing net zero energy buildings.

Understanding automation hardware is not possible without knowing the concept of parallel and series connections. There are two ways of connecting loads to a voltage source; these are shown in Figure 2.4 and Figure 2.5. In a parallel connection, the loads R_1 and R_2 have the same voltage because they are connected to common points from the power source. Even though the voltage is the same, the current through them will be different if the resistances are different. The currents through R_1 and R_2 are calculated as:

$$I_1 = V/R_1$$
$$I_2 = V/R_2$$

The total current passing through the source is the sum of the two currents.

$$I = I_1 + I_2 = V/R_1 + V/R2 = V \text{ x } (1/R_1 + 1/R_2)$$

Therefore, the equivalent resistance of the circuit is

$$R_{eq} = (1/R_1 + 1/R_2)$$

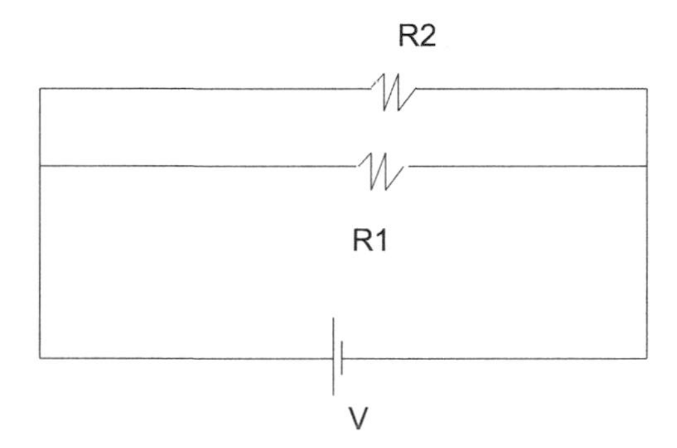

Figure 2.4 Parallel connection

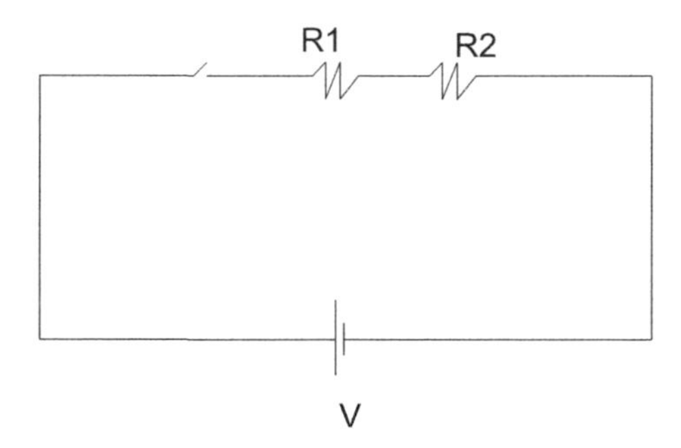

Figure 2.5 Connection in series

In series connection, the same current (I) passes through both the loads R_1 and R_2. But the voltage-drop across them will be different, as given by

$$V_1 = I\,R_1$$
$$V_2 = I\,R_2$$

The total potential-drop across both the loads will be equal to the voltage of the power source. That is,

$$V = V_1 + V_2 = I\,R_1 + I\,R_2 = I\,(R_1 + R_2)$$

Therefore, the equivalent resistance of the circuit R_{eff} is

$$R_{eff} = R_1 + R_2$$

Exercise 2.1 Numerical example – resistors in series

A wire has resistance of 0.1 Ohm per meter. It is used to connect a sensor to a controller that measures the voltage produced by the sensor. The input channel of the controller can be imagined as a voltmeter. The length of the wire is 10 m. The sensor produces an output of 1 Volt, and the current through the wire is 10 mA. What is the voltage read by the voltmeter?

Answer

The resistance of the wire (R1) can be considered as connected in series with the voltmeter as in Figure 2.6. The total resistance of the wire R1 = 10 × 0.1 = 1 Ohm. The potential drop across the resistor R1 is therefore 0.01 V. Instead of reading 1 V produced by the sensor, the voltmeter reads only 0.99 V. There is a 1% error in the reading because of the resistance of the wire.

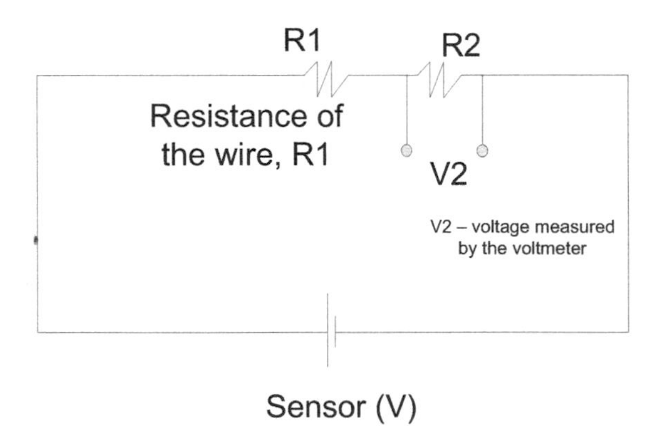

Figure 2.6 Sensor connected to a voltmeter

Exercise 2.2 Numerical example – parallel circuit

USB hubs are used to connect multiple devices to the same USB port of a computer. The devices that are connected to the USB hub form parallel circuits, drawing power from the same USB port of the computer. The USB port of the computer has 5 V potential difference. If two Arduino controllers are connected to a computer through a USB hub, what is the total power supplied by the computer? Assume that each controller draws 0.5 A current.

Answer

The power consumed by each controller = 5 V × 0.5 A = 2.5 W. Since both controllers are connected in parallel, they share the same potential of 5 V. The total current drawn from the computer will be 0.5 A times two, that is, 1A. The total power drawn is 2.5 × 2 = 5 W.

2.3 Essential hardware

2.3.1 Power source

Every circuit needs a power source. Most automation hardware such as sensors and controllers work with low voltage DC. Most of them require a power source having 5 V to 24 V DC potential difference. Since the electricity grid usually provides AC power, we need AC to DC convertors. A DC power adaptor is shown in Figure 2.7. It has pins for connecting wires supplying AC power, and pins for outputting DC power at 24 V. While choosing the power source, we need to consider the voltage, the maximum current, and the maximum power permitted by the adaptor. Wrong adaptors could damage the circuit and the devices. Care should also be taken to connect devices with the correct polarity. Reversing the positive and negative terminals could result in incorrect functioning of the device and damage to it.

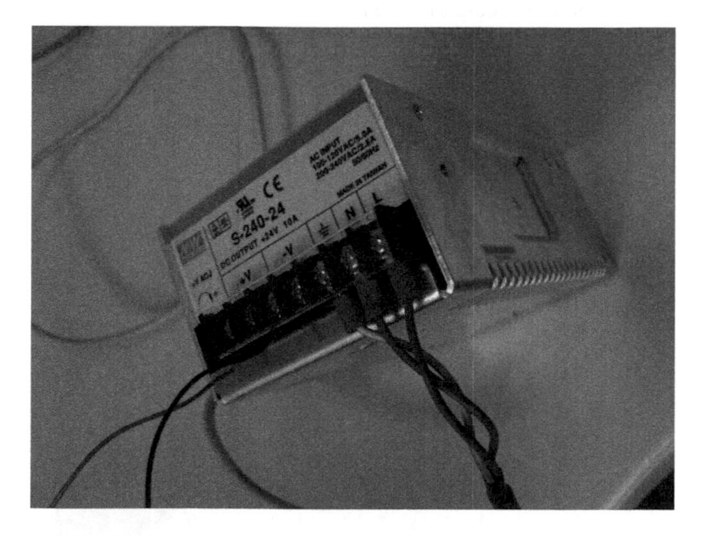

Figure 2.7 DC Power adaptor

2.3.2 Cables

Cables are used for transmitting power as well as data. An AC power cable usually has three wires, for line, neutral, and ground. The ground wire is provided to remove any voltage developed on the body of the equipment by discharging it to the earth. A DC power cable usually has only two wires, one for positive and the other for negative. The wires are color-coded for easy identification of the positive and negative sides. Usually, red is positive, and the negative wire is either black or blue. If the cable is used for transmitting data in the form of voltage or current, it might be necessary to prevent induced currents in the wire due to electromagnetic interference with electrical equipment nearby. Twisted pair cables and shielded cables have been used to achieve this.

Since an automation installation might involve large number of devices, cabling could soon become very messy. Messy cabling could cause lot of confusion and waste significant amount of time fixing problems due to wrong connections. A professionally implemented system is shown in Figure 2.8. All the cables are neatly labeled, and proper connectors are

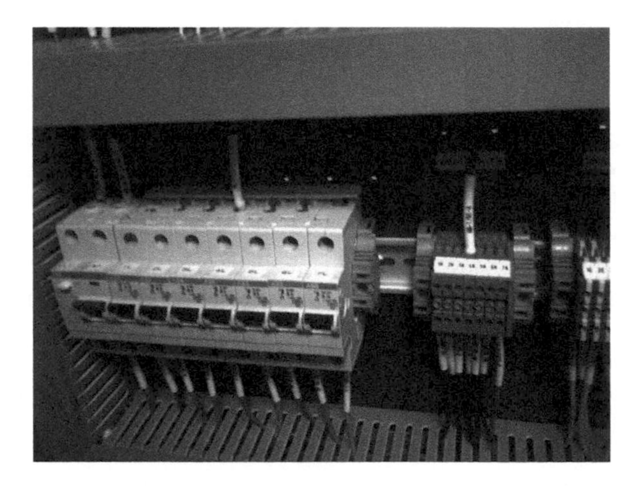

Figure 2.8 A distribution box for a control system

Figure 2.9 An amateur control system in a demonstration project

used to connect the cables such that there is no loose contact. Compare this with an amateur prototype shown in Figure 2.9.

2.4 Sensors

A sensor is any device that measures a physical state. There are many types of sensors used in building applications. They are used for monitoring as well as taking control actions based on the current state. For example, temperature and humidity sensors are used in air conditioning; motion sensors and light sensors (photosensors) are used for controlling lights; chemical sensors that measure the amount of CO_2 and Volatile Organic Compounds (VOC) in the air are used to determine how much fresh air should be supplied.

With the widespread use of sensors in electrical, mechanical, and building systems, sensor interfaces have been standardized. There are three standards:

- Analog output: produces a DC voltage (typically in the range 0–10 V or 0–5 V) or current output (typically in the range 4–20 milliamperes). The voltage or current produced is proportional to the value of the measured parameter.
- Binary output: output two discrete values, high or low voltage.
- Digital output: produce voltage pulses that are interpreted as binary numbers. Typically, two distinct voltage levels are used to denote ones and zeros.

Analog output devices measure a continuous range of values of parameters. The mapping between the voltage and the physical parameter is obtained through calibrating the sensor. Many measurements are taken under controlled conditions in which the actual value of the physical parameter is known. The corresponding output voltage (or current) of the sensor is also recorded. By plotting the points with the voltage on the x-axis and the actual values of the parameter on the y-axis, a curve is fitted to pass through the points as closely as possible. This calibration curve is used to convert the voltage into the physical parameter. Ideally, a linear calibration curve is preferred since the conversion is simple to implement. Many data acquisition systems allow users to input the calibration curve by specifying the slope and intercept of the line. Inputting a non-linear curve is more difficult.

Exercise 2.3 Numerical example – analog output

A temperature sensor produces an output in the range 0–10 V DC. It measures temperature in the range 10°C–90°C. At a certain time, the output voltage measured is 1.5 V. What is the temperature?

Answer

The temperature range of 10°C to 90°C is mapped to voltage levels 0 to 10 V. Therefore, 1 V corresponds to 9°C. Since 0 V corresponds to 10°C, 1.5 V corresponds to 23.5°C.

Digital output devices produce binary values. There are many devices that detect binary states; whether the user has pressed a button, whether there is someone in the room, whether smoke is present in the room, whether a parameter has exceeded permitted values, etc. Such devices need only a binary output, to indicate whether the result is yes or no. Depending on the convention, a zero voltage indicates a no value and a specific voltage (say 5 V) indicates yes value.

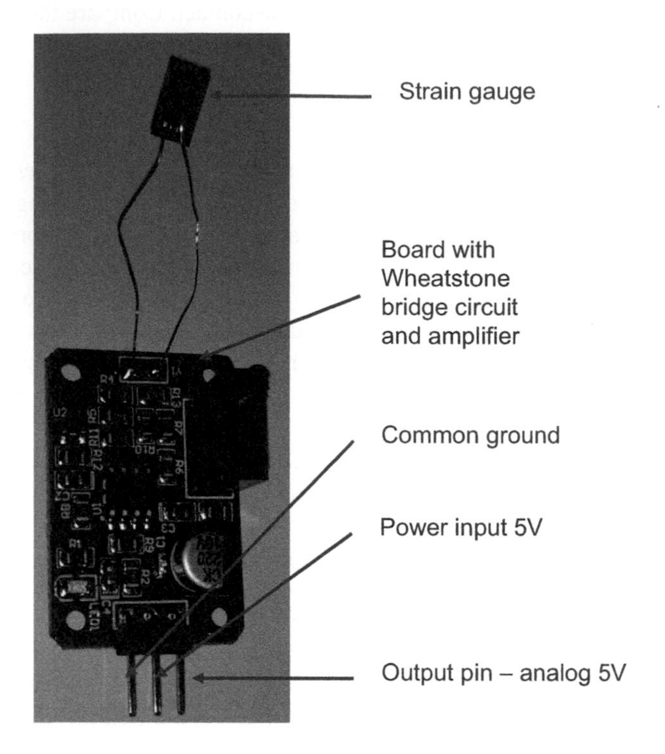

Figure 2.10 Analog voltage output from a strain gauge

Using the binary states of zero and one, more complex data could be transmitted by the sensors. Data are converted into a series of bits representing either one or zero. This is discussed in the section on serial communication.

Sometimes, low-level sensors do not produce standard output in the form of voltage or current. Such sensors need transmitters to convert the output into standard forms. A common example is the output in the form of a change in resistance. Standard strain gauges work based on the principle that the electrical resistance changes with the length of the material. Due to strain, when the length of the material changes, there is a minute change in the electrical resistance. The change in the resistance can be amplified and converted into a current using what is called as a *bridge circuit*. Different types of bridge circuits are available, and these are usually available in data acquisition systems for converting resistance into voltages or currents in the ranges that can be read by reading units. However, if low-level controllers are used, separate transmitters might be needed, and these require bridge circuits to produce the output that can be read by the input channels of controllers. An example of such a board is shown in Figure 2.10. Even though the strain gauge produces output in the form of variable resistance, the output pin of the board produces 5 V analog output that can be read by a microcontroller. Such transmitters can be used for estimating loads in construction automation.

2.5 Actuators

An actuator is a device that performs an action. There are two types of actuators: electrical and mechanical. Electrical actuators regulate currents in circuits. Mechanical actuators apply force or produce motion. Actuators take input in the form of current and voltage, performing the required action. Analogous to sensors, there are three standard interfaces:

- Analog input: DC voltage (typically in the range 0–10 V or 0–5 V) or current (typically in the range 4–20 milliamperes). The voltage or current is proportional to the value of the amount of movement required.
- Binary input: two discrete values, high or low voltage.
- Digital input: voltage pulses that are interpreted as binary numbers. Typically, two distinct voltage levels are used to denote ones and zeros.

Sometimes, the input of the actuator can be directly connected to the output of sensors; the sensor value will then directly determine the amount of actuation. However, it is preferred to use a controller to actuate the output devices instead of sending signals directly from the sensor. Better control of the actions is possible using a controller as shown in Figure 2.11.

2.5.1 Electrical actuators

Electrical actuators are typically used for operating electrical equipment. Manually operated switches have limited applications in automation. We need switches that are operated electronically. Depending on the present state and external conditions, we might want to start or stop a device programmatically. Electrical actuators open or close a circuit or send signals in the form of currents for operating devices.

The most common device for this is a magnetic relay switch. It can be considered as a programmable switch. We are used to manual switches to turn on fans and lights. Suppose we want to do that programmatically through a computer. We need a device that can open or close a circuit by sending the command from a computer. Commands from computers can only be sent using electrical signals. So, we need a switch that operates based on control signals in the form of currents or voltages.

An electromechanical relay contains an electromagnet that attracts a metal piece when a current is passed through the magnet. See Figure 2.12. When a current is applied to the coil, the electromagnetic forces pull the metal piece S causing the circuit to open or close. Thus, the equipment can be started or stopped. The control current needed to activate the coil is small and can be sent through a controller. The voltage and power needed for the equipment could be large; this is not supplied by the controller and is part of a separate circuit with its own power source. The controller provides only a signal to activate the magnetic coil.

A contractor is a special type of relay used in circuits with high loads, such as motors. Contactors are more rugged devices but use the same principle of an electromagnet. A system of relays and contactors used for starting and stopping motors in a coordinated lifting system is shown in Figure 2.13. The components at the center are relays. The ones on the edges are contactors. The relays are sent control currents from a controller. The relays open or close the circuits that power the control currents to the contactors. The contactors start or

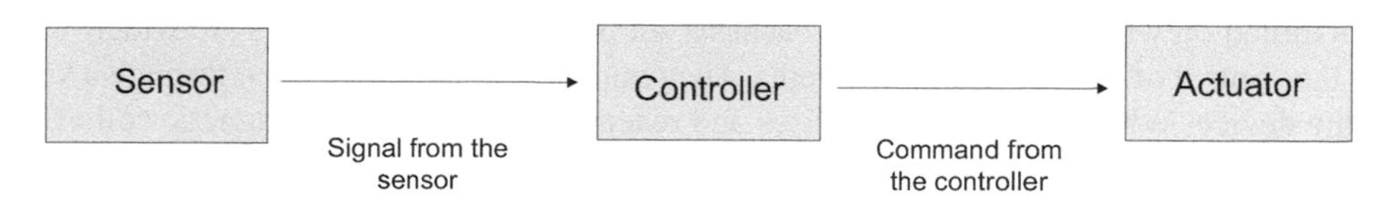

Figure 2.11 Operating an actuator through a controller

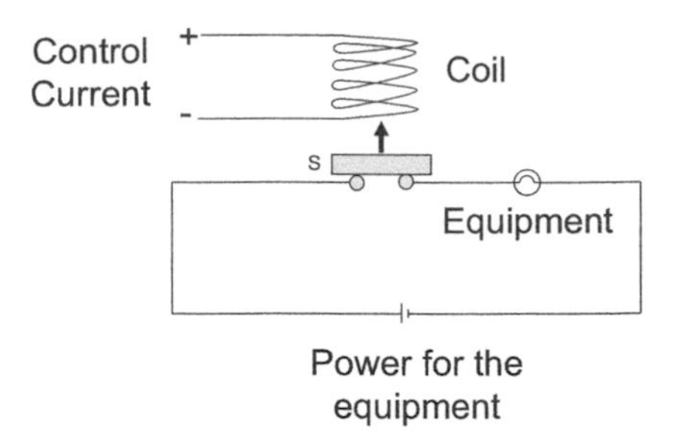

Figure 2.12 Magnetic relay switch

Figure 2.13 Relays and contactors in a coordinate lifting system

stop the motors. The control current required for the contactors is larger than what can be provided by the microcontrollers. Hence, both relays and contactors are used.

The magnetic relay is an electromechanical device with moving parts. The moving parts cause noise and affects the reliability of the switch in the long term. The switch could get jammed and the system might fail, especially when there is high current in the main circuit. Solid state relays have been developed that avoid these problems. These are made of semiconductors, which have no moving parts.

A different type of electrical actuator is a thermostat relay switch. Thermostats make use of the principle that different metals expand at different rates when the temperature is raised. A bimetal strip bends due to unequal expansion. This principle is used to make switches that gets turned on when the temperature reaches a set point. Thermostat relays are widely used for the control of air conditioning systems. The temperature setpoint is input through a knob in the device. When the temperature rises and reaches the setpoint, the magnetic coil of the relay de-energizes. When the temperature falls below the setpoint, it re-energizes. In addition to its use as an actuator, the thermostat usually contains output terminals that are used to transmit the current temperature. Thus, a thermostat is both a sensor and an actuator.

Exercise 2.4 Review questions

1 You want to turn on the fan only when the temperature exceeds a certain value. What actuator can be used for this?
2 Why is a solid-state relay more reliable than an electromechanical relay?

2.5.2 Mechanical actuators

There are many types of mechanical actuators, depending on the type of motion. Rotary actuators, linear actuators, solenoid valves, dampers, etc. The most basic and common actuator is a motor that produces rotary motion. Many other types of actuators use motors for producing other kinds of movement. The torque and speed are the most important parameters while selecting a motor.

There are DC and AC motors. AC motors are typically used for higher loads. AC motors used to pump chilled water in the centralized air conditioning system of a commercial building can be seen in Figure 2.14. Each pump in this system consumes 110 kW of electrical power.

A DC motor works on the principle that a current-carrying conductor experiences a force in a magnetic field. Permanent magnets in a DC motor produce a magnetic field concentrated by an iron core at its center. Coils of wires carrying electric current rotate around the core due to electromagnetic forces that are generated by the movement of electrons in the wires. The torque produced by the coils is transferred to the motor shaft to which the external load is connected.

A small DC motor is shown in Figure 2.15. It has only two wires, one for connecting the positive (top) and the other for the negative terminals (bottom) of a power source. Operating the motor is as simple as connecting the wires to a battery with the correct voltage.

The speed of a DC motor depends on the applied voltage; for a constant load, you can increase the speed by increasing the voltage up to the maximum permitted value. The speed is inversely proportional to the load (torque) for the same voltage. The maximum speed occurs when there is no load, and the speed decreases linearly with the load. The maximum speed of the motor is specified in rotations per minute (RPM). Gears could be attached to

Figure 2.14 Chilled water pumps in a plant room

Figure 2.15 DC motor

the shaft to adjust the speed of rotation. Changing polarity causes rotation in the reverse direction.

Motors should be selected such that the technical specifications meet the requirements of the application. Most crucially, we must estimate the load or the torque to be taken by the motor and the speed required. An example of the specifications of a DC motor follows:

- Speed 60 RPM
- 12V DC supply
- 6mm shaft diameter with threaded hole and screw
- 2 kg cm torque
- No-load current = 60 mA (max), load current = 250 mA (max)
- Motor with gear

The simple DC motor does not provide any feedback on the actual number of rotations made. This information is needed for accurate control of the movement. Stepper motors and servo motors make use of this information.

Servomotors

Servomotors are AC or DC motors with position feedback. Encoders are used to measure how much the motor shaft has rotated. Shaft rotation position can be determined by a potentiometer or using optical sensors. The resistance of the potentiometer changes when the shaft rotates. By measuring the resistance, the angle of rotation can be determined. Optical encoders operate by counting markers with the use of a light source and a photodetector. Information from the encoders is used for precisely controlling the rotation of the servomotor.

A small DC servomotor is shown in Figure 2.16. From outside it looks like a simple DC motor. The encoder and other details are not visible from outside. However, comparing with Figure 2.15, you will notice that it has more wires. Additional wires are needed to provide inputs for controlling the servo.

In some servomotors, the rotation of the shaft is restricted to angles within a certain range, for example from 0° to 270°. There are continuous rotation servomotors without this restriction. Since in continuous rotation servos there is no limit to the angle (it can be several multiples of pi), the speed of rotation is usually controlled instead of the angle.

A piece of electronic hardware in the form of a board called a servo controller is used to operate a servomotor. The angle or speed is provided as input to the servo controller. A standard method of providing input to a servo controller is using a sequence of voltage pulses, called as a pulse train (Figure 2.17). By convention, the width of the pulse indicates the

Figure 2.16 DC servomotor

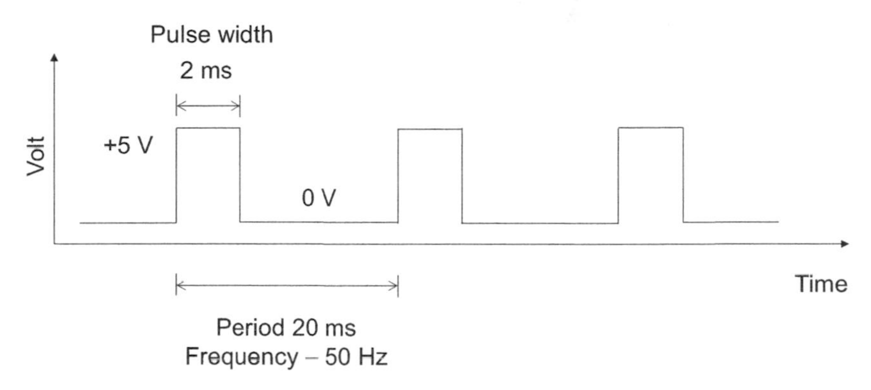

Figure 2.17 A pulse train to control a servo

required position. If the required position is different from the current position, the controller will provide a voltage to the motor to move forward or backward. The voltage is adjusted until the required position is attained. Standard servo controllers use 50 Hz pulses (pulses are repeated 50 times a second, that is, one pulse every 20 milliseconds). The pulse width is usually in the range 1 to 2 milliseconds, with the lowest pulse width indicating the minimum position and the maximum pulse width the maximum position. In the case of continuous rotation servos, instead of rotating the shaft to a particular angular position, we specify the speed of rotation in the form of the pulse width. The controller tries to make the motor rotate continuously in the specified direction at the specified speed.

Microcontrollers can generate pulse trains through their output channels by setting the output to an *ON* position (high voltage) for the required time interval. Functions to generate pulses of specified widths and frequencies are part of the software development libraries of popular controllers. Thus, we can programmatically change the speed or position of servomotors through these controllers.

Stepper motors

Stepper motors are DC motors that have multiple groups of coils. The motor rotates by a small angle when a coil is energized. By repeating the process for each group of coils, the motor rotates in discrete steps, one at a time. Precise control of the position can be achieved by stepper motors. Hence, they are used in 3D printers and robotics. However, since they typically work on low voltage DC, the torque and power are limited.

Dampers

Dampers are common in air conditioning applications. They are used to control the flow of air through ducts by opening or closing the valve. In Figure 2.18, a damper along with its controller is shown. At the top of the damper, there is a clamp to hold the shaft of the valve. By rotating the shaft, the valve opens or closes, thus changing the size of the opening of the duct. The controller rotates the motor of the damper such that required opening is achieved. A damper installed on a real air conditioning duct is shown in Figure 2.19.

Motorized dampers help to automate the air flow in air conditioning ducts in response to changing conditions. In some systems, dampers are manually operated. An example is

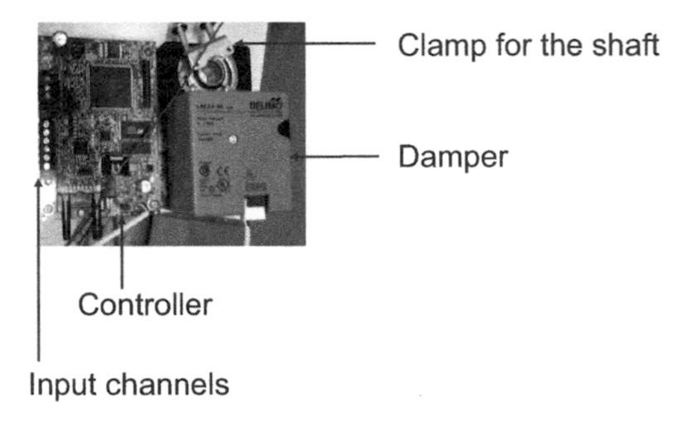

Figure 2.18 A damper

Figure 2.19 A damper installed in an air conditioning duct

shown in Figure 2.20. By manually turning the knob of the handle, the valve can be opened or closed (Figure 2.21). Manual operation of the damper is simple and easy to implement where the air flow need not be changed very often. However, the use of motorized dampers allows much flexibility and results in smart autonomous behavior. Motorized dampers are easily created by connecting the shaft of the valve to a motor using a gear.

Solenoid valves

Solenoid valves are installed in pipes transporting liquids, and they perform the same function as a damper in an air duct. It opens or closes the pipe to control the flow through it and works by magnetizing an electric coil that pulls a magnetic core called the plunger. As the plunger moves, the pipe is opened or closed. This device could be used in water supply systems and automated irrigation systems.

Linear actuators

Linear actuators produce linear motion instead of rotary motion. Linear motion is needed in many automation systems. It is easier to work with rectilinear coordinate systems. Specifying

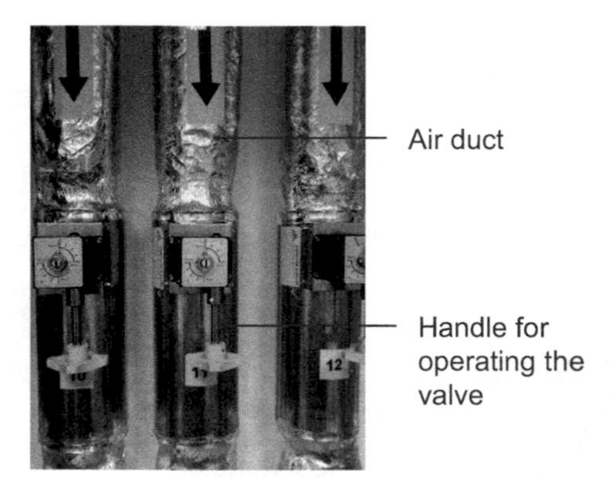

Figure 2.20 Manual dampers in air ducts

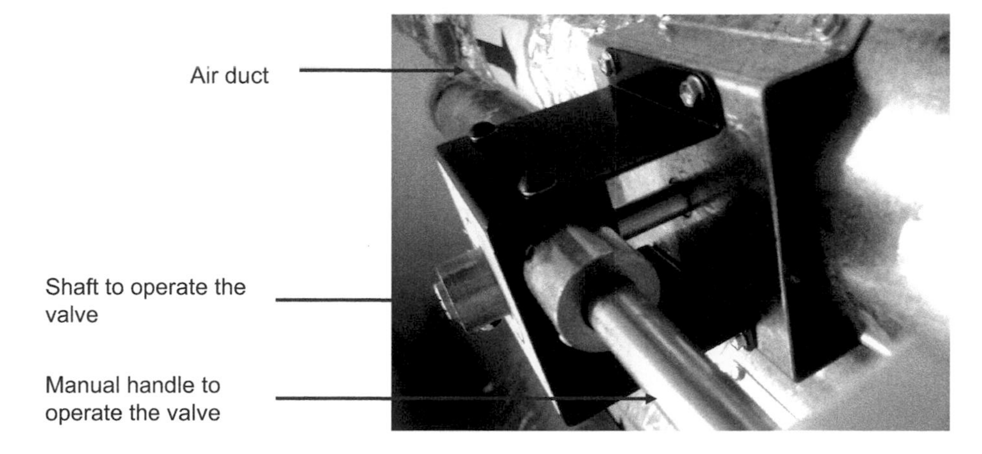

Figure 2.21 Manually operating the damper with a handle

linear movements in the x, y, and z directions is convenient for accurately positioning an object in the workspace. Linear actuators are also used to apply a force on an object in a specific direction. For example, a pin can be inserted into a hole for connecting two parts by pushing the pin with a linear actuator.

In some linear actuators, motors are used to produce rotary motion, which is converted into linear motion using special arrangements such as lead screws, ball screws, worm screws, and belts. A typical arrangement is shown in Figure 2.22. A motor rotates a threaded rod called a *lead screw*, using gears. There is a nut which normally moves through the threads of the lead screw when rotated. When the nut is connected to the load by a welded plate, it is prevented from rotating. When the rod is rotated, the nut is pushed up or down by the forces in the thread. Therefore, it moves linearly; it moves up and down when the rod rotates, thus lifting or lowering the load.

Worm screw jacks operate in a similar manner. The rotary motion of the motor is used to rotate a lead screw using specially designed gears. Rotation of the lead screw results in linear motion of the load at the top of the rod. A worm screw jack is shown in Figure 2.23. A system for lifting loads vertically using worm screw jack is shown in Figure 2.24. The motor is engaged with the gears of the worm screw jack. The load is connected to the top of lead screw, mounted on bearings to prevent the load from rotating when the screw turns.

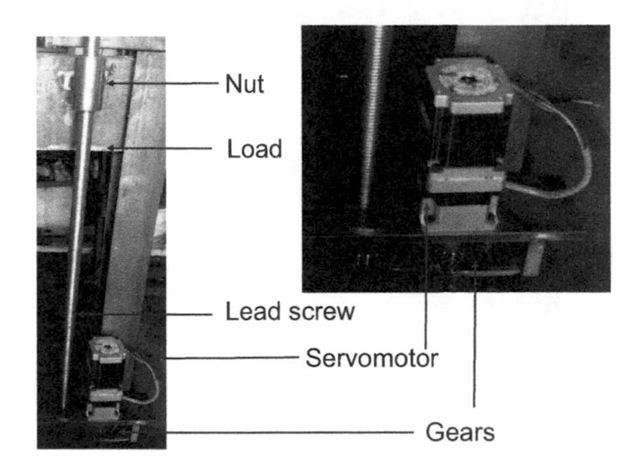

Figure 2.22 A linear actuator using lead screws

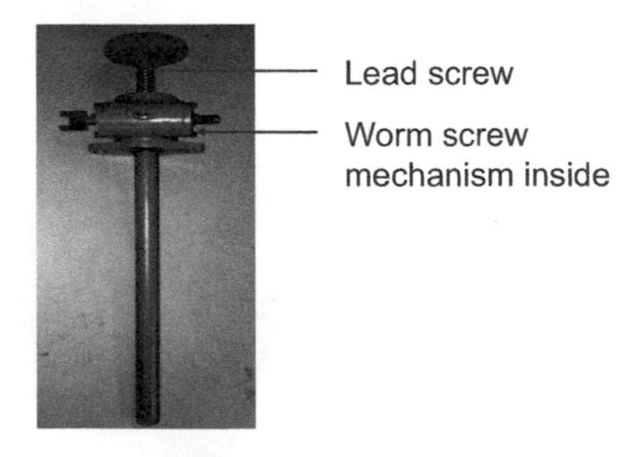

Figure 2.23 A worm screw jack

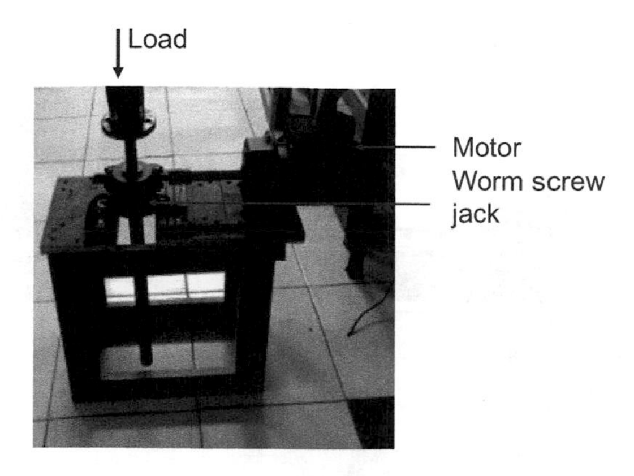

Figure 2.24 Vertical lifting using a worm screw jack

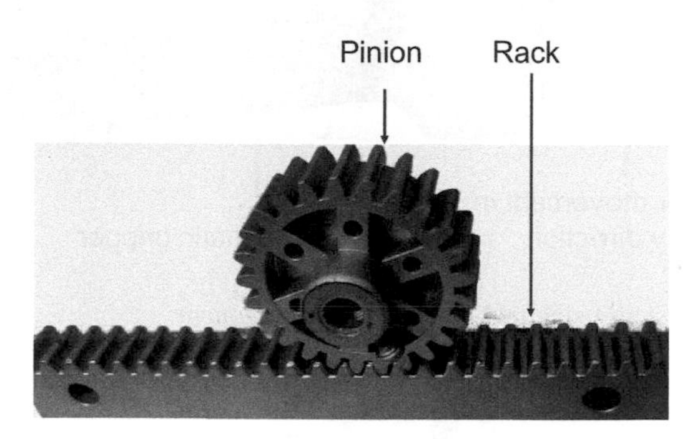

Figure 2.25 Rack and pinion

Rack and pinion

A rack-and-pinion system consists of two main parts, a circular gear called a pinion rotating on top of a linear gear called the rack (Figure 2.25). A motor is used to rotate the pinion, and since the teeth of the pinion are engaged with those of the rack, it moves forward. The rotary motion of the motor is thus converted into linear motion of the pinion. This mechanism has been used to produce x-y movements in 3D printers. A model of a gantry system for picking and placing reinforcement bars is shown in Figure 2.26. The x and y movements are achieved through racks and pinions.

Hydraulic and pneumatic actuators

Hydraulic actuators produce linear motion through applying fluid pressure in a piston. The piston is inside a cylinder that is connected to a system that pumps fluid, typically oil (Figure 2.27). The pump is typically powered by an electric motor. A valve is located between the cylinder and the reservoir containing the fluid. This is used to control the pressure in the cylinder.

Hydraulic systems are usually more efficient in transmitting forces than gears and axles. The same pump can be used to move multiple cylinders that are oriented in different

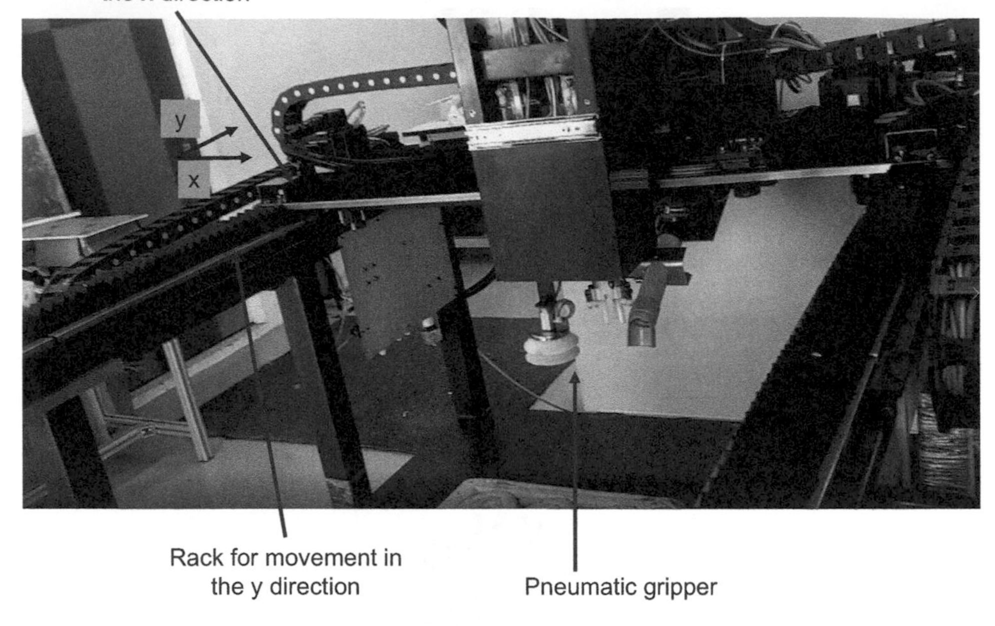

Figure 2.26 A gantry system using rack-and-pinion arrangement

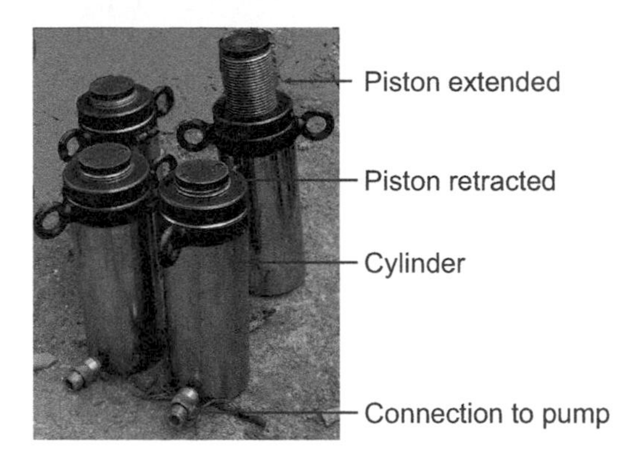

Figure 2.27 Hydraulic jacks

directions without complex mechanical systems for changing the direction of the force. Furthermore, it can transmit large forces. Hence these are the preferred means of applying forces in large construction systems.

The use of hydraulic jacks is demonstrated using examples in metro rail construction. A launching girder is a piece of equipment used in the construction of viaducts for metro rails. It is used for lifting precast concrete segments. The launching girder is initially supported on the piers of already completed spans (See Figure 2.28). In this launching girder, there are three groups of legs to support the load of the girder, one at the front, one in the middle, and one at the rear. The girder is pushed to the next span to assemble the viaduct there; this

Figure 2.28 A launching girder supported on an already constructed span of metro rail

Figure 2.29 Launching operation

operation is called launching. While launching, the front legs of the girder are lifted so that they lose contact with the supports and do not take loads. During this time, the middle and rear legs support the girder and the front portion cantilevers out of the already constructed span (see Figure 2.29). At the end of the launching operation, the front part of the girder is over the supporting pier of the next span that is yet to be constructed. At this point, the front legs come down again and make contact with the support. After the launching girder is firmly supported on the piers, it starts lifting the precast concrete segments for constructing the next span of the viaduct (Figure 2.30).

Hydraulic jacks are used for lifting and lowering of the legs of the launching girder at the start and end of the launching operation. Hydraulic jacks used for the supporting legs of a launching girder is shown in Figure 2.31.

The launching operation also makes use of hydraulic jacks. Push-pull jacks are used for this. It works like this: The base of the jack is first bolted to rails with the piston fully retracted. The

Figure 2.30 Lifting of segments. The segments are pulled up using cables connected to the launching girder above.

Figure 2.31 Hydraulic jacks for the supporting legs of a launching girder used for constructing metro viaducts

tip of the piston is connected to the launching girder to be moved. Then the piston is pushed out by operating the pump. As the piston comes out, it pushes the girder, and it moves forward. After the full-stroke length of the piston is reached, the base of the jack is unbolted from the rails, and the piston is retracted. Now, it is connected to the next bolt hole on the rail, and the process is repeated. A hydraulic jack used for launching of the girder is shown in Figure 2.32.

After the launching operation, the girder is supported on the piers of the span that is yet to be constructed. It starts lifting viaduct segments made of precast concrete. The

Figure 2.32 Push-pull jack for the launching operation of a launching girder

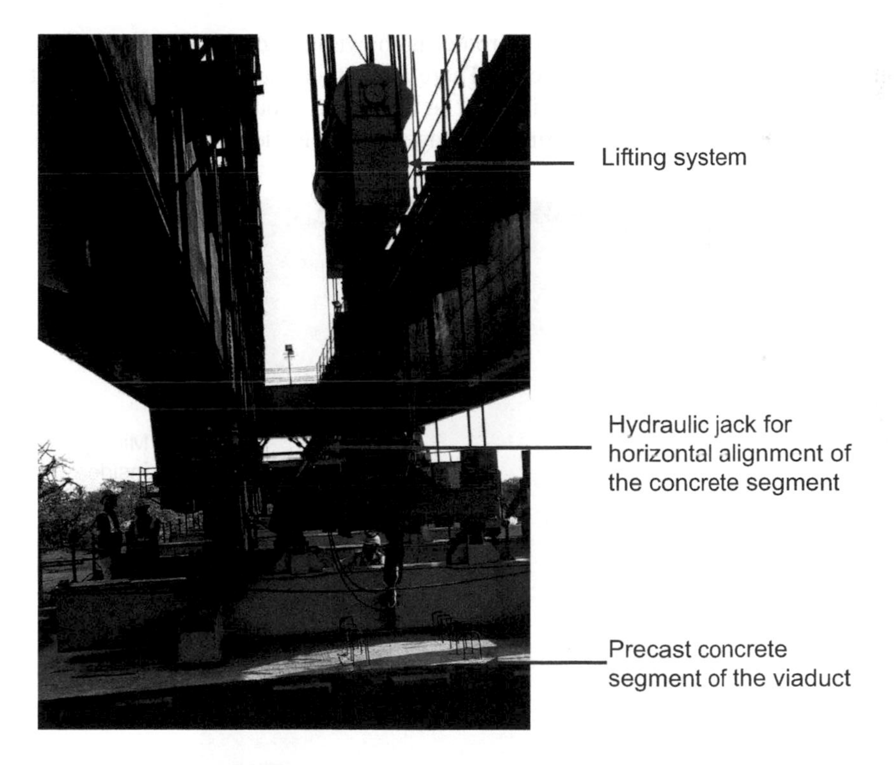

Figure 2.33 Hydraulic jack for alignment

segments need to be aligned horizontally before they are connected together. A hydraulic jack for this alignment is shown in Figure 2.33. By extending and retracting this jack, the precast segment can be rotated such that it is aligned perfectly with the adjacent segment.

2.6 Controllers

The term controller is used to denote both software and hardware. In this book, the term is used to denote the hardware that performs three functions:

- Read data from sensors (data acquisition)
- Perform computations to determine appropriate actions to be taken
- Send commands to actuators

Three commonly used types of controllers are

- Direct Digital Controllers (DDC)
- Programmable Logic Controllers (PLC)
- Microcontrollers

DDCs are widely used in building automation applications. PLCs are more commonly used in industrial systems. Microcontrollers are more popular in robotics applications and hobby electronics projects. This book mostly uses microcontrollers as examples for illustrating automation applications.

A direct digital controller with inbuilt router is shown in Figure 2.34. It has a set of input channels to which sensors are connected. There are output channels to which actuators are connected. It has a microprocessor like that in desktop computers. Programs can be written in popular high-level programming languages and uploaded to the microprocessor for

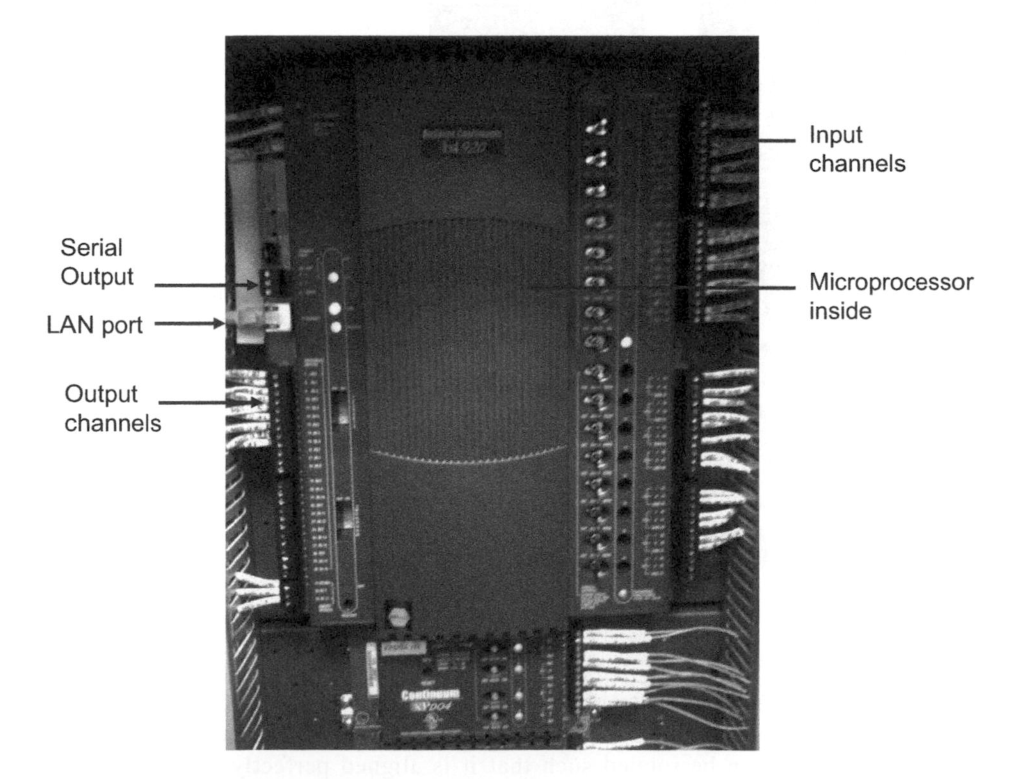

Figure 2.34 A direct digital controller with inbuilt router

performing computations to analyze the input data and produce output to be sent to actuators and other devices. This DDC has an inbuilt router to communicate with a local area network (LAN). There is a LAN port through which the controller can be connected to a network switch, and other computers in the network can send messages to it using TCP/IP protocol. In addition, there is a serial port through which other controllers can communicate with it using BACNET protocol (See Section 3.2.4).

PLCs also have input and output channels and perform the same functions as a DDC. However, the programming methodology is a bit different. A PLC program consists of a set of instructions containing the application logic for controlling real-time systems. It is common to use graphical languages such as ladder diagrams, which are convenient for representing simple logic. Modern high-level programming languages are rarely available for PLCs. The emphasis is on real-time performance rather than flexibility in programming.

Open-source microcontrollers have entered the market more recently, after hardware became cheap. A popular microcontroller called Arduino Uno is shown in Figure 2.35. This controller is 100 times cheaper than the one shown in Figure 2.34. However, they perform similar functions. Arduino has a few analog input pins and several digital input/output pins which can be used to connect sensors and actuators. This microcontroller does not have built-in support for protocols such as BACNET. It does not natively support TCP/IP protocol and does not have a LAN port. It does not even have permanent storage in which user data could be stored during the execution of the program. It only has a limited amount of flash memory for uploading the program. It communicates primarily through the USB port and other serial ports. However, Ethernet connectivity, storage, and other features could be added through external hardware connected to the input/output pins of the controller. The microcontroller is not expensive because of limited features and because it does not depend on proprietary hardware or software. The design is kept as simple as possible with minimum functionality. Even though these are mostly used for hobby electronics and prototype development, it has been successfully deployed in commercial products.

The popularity of low-cost microcontrollers like Arduino prompted large companies to develop similar products. An example is the Intel Galileo board (Figure 2.36), which has similar features like Arduino, but with a more powerful processor and more memory. Even though, this product was discontinued after some time, several others have sprung up.

Raspberry Pi was introduced as a small computer that was affordable to the masses (Figure 2.37). It is now being used as a microcontroller in many automation projects. It is driven by modern open-source operating systems with powerful capabilities. It has persistent storage (SD card), and Ethernet and Wi-Fi connectivity. The processor is more powerful than an Arduino and has a real-time clock. The original Raspberry Pi did not have analog input pins and was purely a digital computer. Then a cheap and tiny version of the hardware called a Raspberry Pi Pico was introduced (Figure 2.38). When it was first introduced, it costed less than 4 US dollars per piece. It has analog and digital pins that can be used to read data from different types of sensors. There is a limited amount of flash memory that is available to store user programs data, but it has no native support for Ethernet connectivity. The emphasis had been to reduce the cost by removing features that are not so essential in a dedicated controller for real-time applications.

Raspberry Pi with wireless connectivity is attractive for developing mobile robots. Its small size, low power consumption, and ability to communicate without wires that obstruct movement make it convenient for mounting on moving robots. This onboard computer can receive data from sensors, perform computations, and send commands to actuators. It can

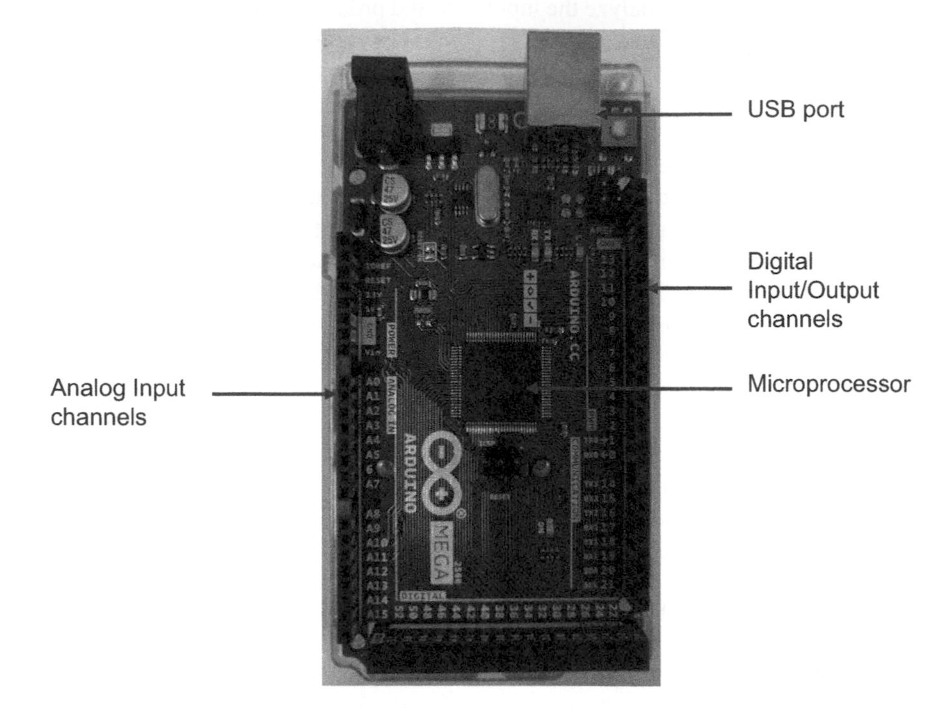

Figure 2.35 An Arduino microcontroller

Figure 2.36 Intel Galileo board

also receive commands from human operators and other devices wirelessly and respond appropriately.

Due to the proliferation of many low-cost controllers, it is now easy to develop demonstration projects without requiring deep knowledge of electronics hardware. New ideas can easily be demonstrated by connecting controllers, sensors, and actuators, and writing simple programs.

Figure 2.37 Raspberry Pi

Figure 2.38 Raspberry Pi Pico connected to an infrared sensor

2.7 Communication hardware

The communication network is an essential component of any automation system. Communication is needed between

a Sensors and controllers
b Controllers and actuators
c Controllers and other devices in the network

The simplest means of communication between sensors, actuators, and controllers is through analog input/output. The sensor produces a voltage or current that is proportional to the parameter to be transmitted. Then the controller reads the voltage or current and converts it into the parameter using the calibration equation. Similarly, the controller sets a voltage proportional to the degree of actuation required, and the actuator reads this voltage and performs the action. Since controllers are not analog devices, the analog voltage that is read by the controller needs to be converted into a digital form and stored in the memory of the controller. This is performed by a circuit called the analog to digital converter (ADC). The number of bits used by the ADC determines the precision with which the numbers could be represented. If a 10-bit ADC is used, only 1,024 (2^{10}) discrete values of the voltage can be read by the controller. Then, the precision with which the physical parameter can be measured will be the maximum range divided by 1,024.

The drawback of this method is the loss of accuracy when long wires are used to connect the devices because of the potential drop due to the resistance of the wire or loose connections. Digital communication using discrete voltage levels provides higher reliability because a high drop in the voltage levels due to losses can be detected, and transmission of wrong data can be prevented.

2.7.1 Digital communication

Digital communication takes place by sending a series of voltage pulses; these are interpreted using a common understanding of the format. The format is defined by the communication protocol. Two different voltage levels are used to represent a one and a zero. The voltage is kept steady for a certain duration according to the agreed rate of transmission, known as the baud rate. This scheme is commonly known as serial communication since bits of data are transmitted serially (one after the other) through the same wire. An example is shown in Figure 2.39. The voltage levels used to represent ones and zeros depend on the hardware specification of the protocol. In older systems, it used to be ±13 V. Today, lower levels in the range 0–5 V are more common.

Serial communication is explained here with an example involving devices that we use in everyday life. Two individuals *A* and *B* living in neighboring buildings wish to communicate using a pair of wires (Figure 2.40). *A* connects a battery to the wires on his side through a switch. *B* connects an electric bulb on her side. When *A* presses the switch, the bulb glows on the side of *B*.

Using this system, *A* can transmit one bit of information to *B*. For example, if both *A* and *B* agree that a live circuit means "Yes", then this information can be transmitted by pressing the switch for some time. However, the scheme is too restrictive in its capabilities. It cannot transmit the information, "No". If the bulb does not glow, it could mean that the battery is down or there is something wrong with the circuit. To reliably transmit both Yes and No, they need to

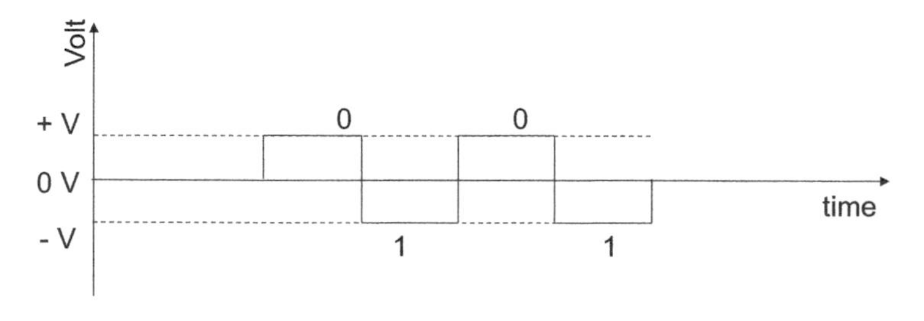

Figure 2.39 Illustration of serial communication

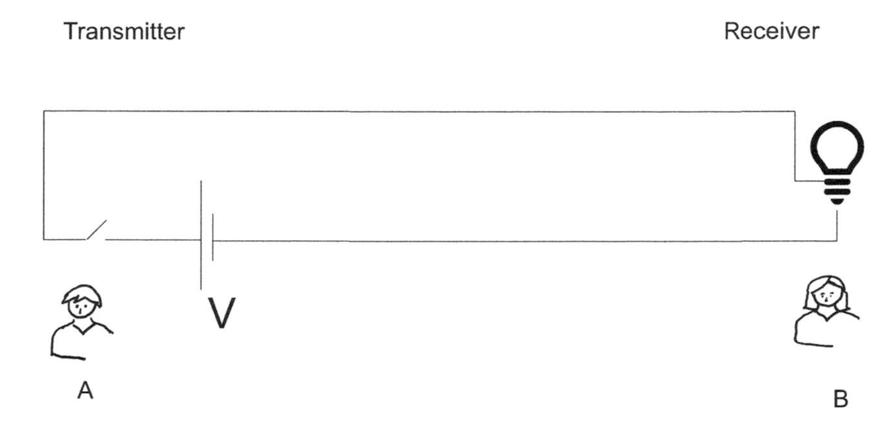

Figure 2.40 Communication using a pair of wires

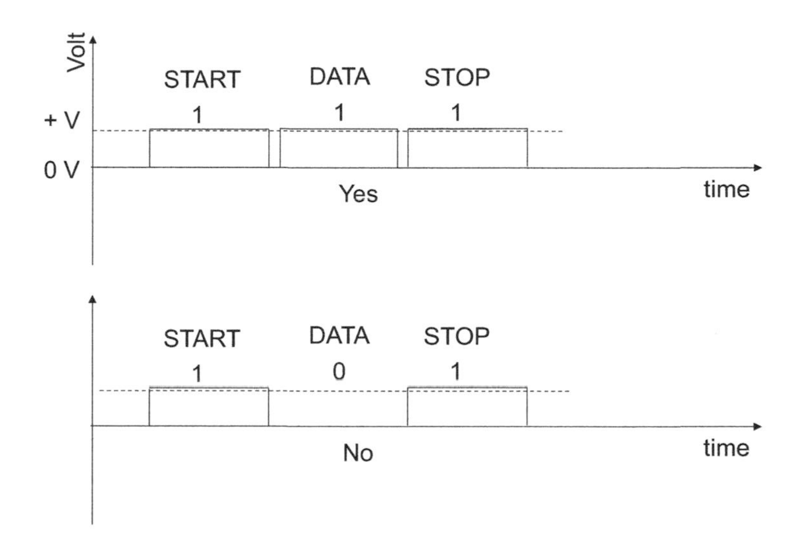

Figure 2.41 Sending Yes and No through a series of pulses

agree on a protocol. For example, they might decide like this: To start transmitting any information, first press the switch for five seconds. Then keep the switch pressed for the next five seconds if "Yes" needs to be transmitted; otherwise, turn off the switch for five seconds. After either outcome, press the switch for five seconds to indicate that the transmission has finished. By observing the bulb glowing for 15 seconds, person *B* is able to decipher the information. See Figure 2.41. Three bits of information are sent. The first part is called the START bit, then the data bit (Yes or No), finally, the STOP bit. The START and STOP bits indicate the beginning and end of the information for the current interpretation of the data bit.

Serial communication makes use of a scheme like the one in Figure 2.41. Instead of a switch and a battery on one side, a semiconductor device is used to switch on the circuit and apply the voltage. Instead of the bulb on the other side, a device to read the voltage is used. The first device is called a transmitter and the second one is called the receiver. The transmitter applies the correct voltage levels sequentially corresponding to the data to be sent. The receiver reads the voltage levels and interprets the data. The sequence of bits, whether there is a start and a stop bit, how many data bits are used, etc. are specified in the communication protocol.

2.7.2 Hardware for serial communication

Hardware for serial communication have been developed, and they are known by different names: SCI (Serial communication interface) and UART (Universal Asynchronous Receiver-Transmitter). RS232 (serial port in older computers) is an example of an implementation of UART. RS232 cables with 9 pins were used to connect peripheral devices to the serial port of a computer. Today, RS232 is replaced by USB (Universal Serial Bus), which has a similar interface for serial communication.

The speed of transmission depends on the hardware used; the speed at which transmitters can switch the values one and zero; the speed at which the receivers can read the bits and process them; and the speed at which the wires can transmit the voltage without distortion in the shape of the pulse. The signals will get distorted if the voltage levels are changed too quickly because of the inductance of the wires – wires store energy and it takes time to drain out the stored energy. In standard serial communication, typical baud rates are 9.8 kbps or 19.6 kbps.

Hardware specifications have been developed to permit higher transmission speeds. Ethernet (using LAN ports) and USB are now the default communication interfaces. Some of the modern communication interfaces require special hardware and software, which may not be available in low-level microcontrollers.

In the illustration shown in Figure 2.39, the width of the pulse, which determines the transmission speed, is fixed a priori. The receiver needs to know the baud rate so that it can sample the input signals at the correct intervals and interpret the pulses as ones and zeros. This is the case when serial communication devices are connected to computers. The baud rate must be specified; only then the messages are read correctly. Both the sender and receiver have to use the same baud rate. These restrictions can be eliminated by using an additional line for sending timing pulses to synchronize the reading and writing of the data bits. This technique is used in protocols such as SPI (Serial Peripheral Interface) and I2C (Inter-integrated chip). Some of the common communication interfaces are briefly introduced in the following sections.

RS232

RS232 typically uses voltage levels −13 V to +13 V. Positive voltage in the range 3 to 13 V is used to denote bit value of 0; negative voltage denotes bit value of 1. It uses either a 9-pin or a 22-pin connector. It is used for point-to-point communication, that is, for exchanging data between two devices. It cannot be used for connecting more devices to form a complex network. Since it uses higher voltage levels, the power consumption tends to be higher.

The pins of an RS232 interface are shown in Figure 2.42. There are separate pins for sending data and receiving data (pins 3 and 2). This means that the two devices can send and receive at the same time (using separate wires). Pin 5 is the common return (ground) for both sending and receiving. The remaining pins are for various other purposes and have origins in the modem communication system of the olden times.

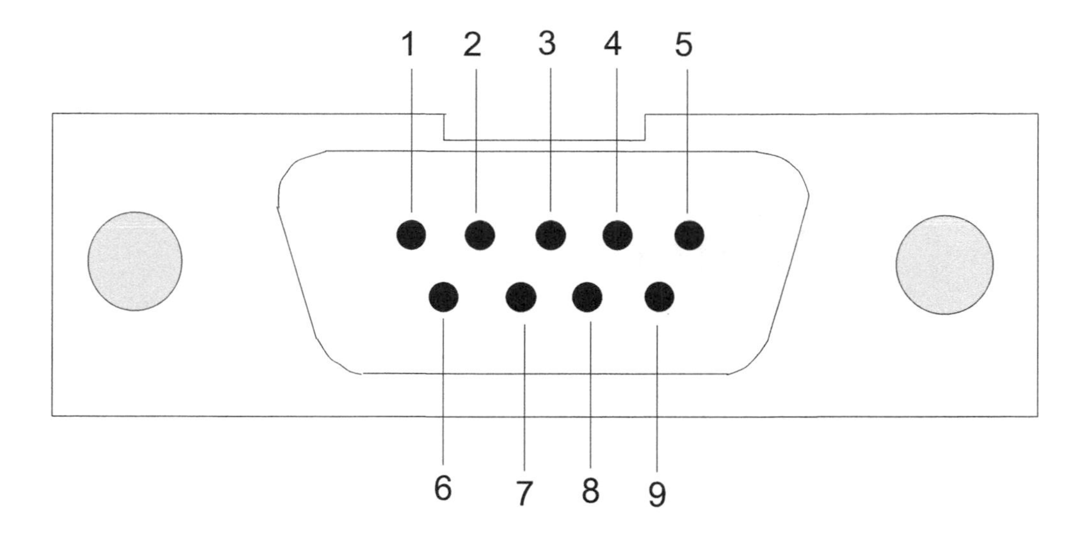

1. Data carrier detect
2. Receive data
3. Transmit data
4. Data terminal ready
5. Signal ground
6. Data set ready
7. Request to send
8. Clear to send
9. Ring indicator

Figure 2.42 RS232 pinout

Ethernet

Ethernet (IEEE 802.3), the standard computer networking protocol today, is also a serial communication standard. The pins of a CAT5 cable used to connect a computer to a LAN port are shown in Figure 2.43. This is called an RJ45 connector. Data are transmitted and received using separate pins marked as TX+ and RX+. Unlike in RS232, there is no common ground pin; there is separate return for transmission and receiving. The Ethernet specification permits much higher transmission speed compared to RS232.

UART TTL

UART is a piece of hardware used to implement serial communication (including RS232). The UART chip converts the data into serial form for transmission through the serial port. TTL stands for transistor-to-transistor logic. It is similar to RS232, but the voltage level is lower; 0–5 V is used. 0 V stands for bit value of 0, and 5 V stands for bit value of 1. Note that it does not use negative voltage, unlike in RS232.

The USB port on a modern computers has four pins: 1) power; 2) data return; 3) data +; and 4) ground. See Figure 2.44 for a picture of the USB connector with the pins exposed. The pins 2 and 3 are used for serial data transmission using TTL. There is simple hardware that can be used to connect a serial port to USB; this converter can be used to send data from a peripheral device having a serial port to a computer with USB port. See Figure 2.45. There are two pins labeled TX and RX in this device. These are for transmitting and receiving data. These pins are connected to the corresponding pins of the peripheral device and the data can be read through the USB port of the computer.

Figure 2.43 LAN cable connector

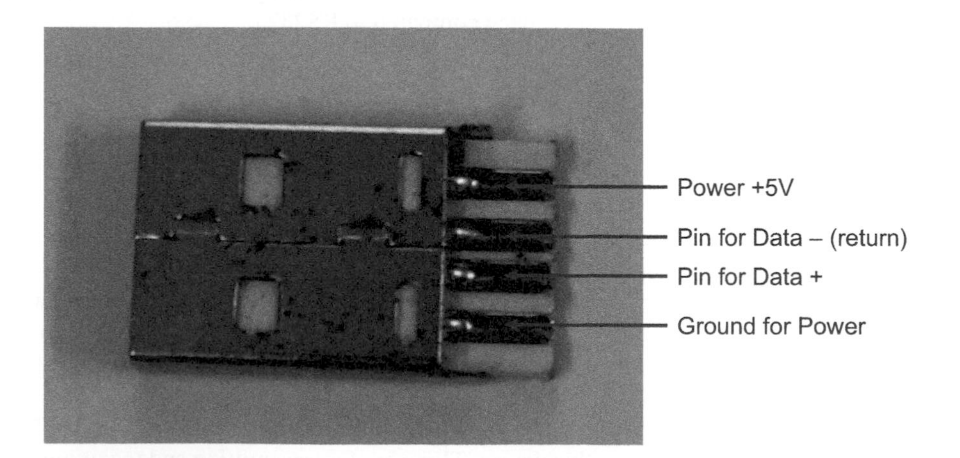

Figure 2.44 USB connector

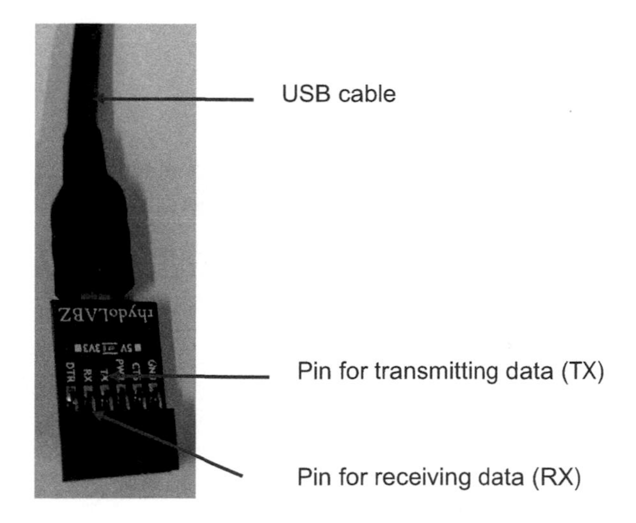

Figure 2.45 USB to UART TTL converter

2.7.3 Routers

Routers are network devices that are used to connect two or more networks. Data packets form one network are transmitted to the second network if the source and destination devices are in different networks. Routers also perform translation of messages to the right language (protocol) used by the networks. For example, BACNET is a protocol used in building automation and control (See Section 3.2.4). BACNET-IP routers translate TCP/IP packets sent from computers through the LAN into BACNET messages that can be understood by BACNET controllers. This makes it possible to send commands from a computer to the controller for reading sensor values or for activating an actuator.

A BACNET router is shown in Figure 2.46. It has a LAN port through which it can be connected to a network switch to communicate with computers on the LAN. In addition, it has an RS485 port which is used to connect it with the controllers in a BACNET network. (RS485 is a serial communication system using just a pair of wires.) A complex network of BACNET controllers and other devices such as IP cameras can be created using this router.

2.7.4 Wireless communication

In principle, wireless communication is not much different from communication using wires. Wireless communication makes use of electromagnetic waves for transmitting information. Antennas transmit and receive waves that encode data. Radio frequency waves are commonly used. Wi-Fi, Zigbee, and Bluetooth are standards that are popularly used in automation systems. Wireless adapters are available that take serial input through pins and transmit data through radio waves. A wireless communication system that was developed as part of a research project is shown in Figure 2.47 (Soman et al., 2017). It uses a Zigbee device that has a range of 100 m and a data transfer rate of 250 kbps. There are pins to connect sensors to the board, and the data from the sensors is automatically transmitted wirelessly to a remote server at regular intervals through a gateway device.

Figure 2.46 BACNET router

Figure 2.47 A wireless communication module (Soman et al., 2017)

2.8 High-level devices

The actuators that were discussed in Section 2.6 were mostly low-level hardware. These are needed for making more sophisticated hardware that permit automation at a higher level. High-level hardware is discussed in this section.

2.8.1 Industrial robots

Different types of industrial robots are now commercially available. These are widely adopted in industries such as automobile manufacturing. They perform tasks such as picking and placing objects at precise locations. Internally, they use stepper or servomotors for creating the motion. However, higher-level interfaces are available for programming the robots such that low-level control of motors is not required. Ideally, the user should be able to input the cartesian coordinates x, y, and z of the trajectory, and the robotic arm should be able to move along the trajectory to perform tasks. In reality, it involves many challenges such as solving inverse kinematics problems (finding out what angles the motors should rotate such that the desired cartesian coordinates are reached), optimization to determine the optimal path, etc.

A 6-DOF robotic arm is shown in Figure 2.48. It has a horizontal reach of 0.5 m, vertical reach of 0.4 m, and a position accuracy of 1 mm. The arm has a pneumatic gripper and can lift a payload of 2.5 kg. The joints are actuated using DC servomotors with gears. Small objects such as reinforcement rods can be repeatedly picked and placed using this robot. However, locating the objects to be picked up requires sophisticated software using machine learning. Planning the path of the arm might also require complex algorithms if the work-space is cluttered with lot of objects.

2.8.2 3D printers

3D printers that can print buildings have attracted lot of attention recently. The idea that a building can be digitally fabricated using a computer model has fascinated a lot of people. Many problems in conventional construction using concrete can be potentially eliminated through 3D printing. For example, formwork can be eliminated, and the time and cost for these activities can be reduced.

The most popular form of concrete 3D printing is through a process called contour crafting (Khoshnevis and Hwang, 2006). It involves depositing material (concrete) layer by layer from top to bottom. It is usually implemented using the concept of a gantry robot in which the printer head is mounted on an overhead frame that moves along the horizontal plane. The x and y movements of the system might be achieved using any of the linear actuation schemes that were discussed earlier. In addition, the printer head can also be lifted and lowered (z movement) using linear actuators. Several other mechanical components are needed for the full functioning of the 3D printer. A concrete pump is needed to extrude the material at specified velocities. The nozzle might have to be rotated to get the correct shape at corners.

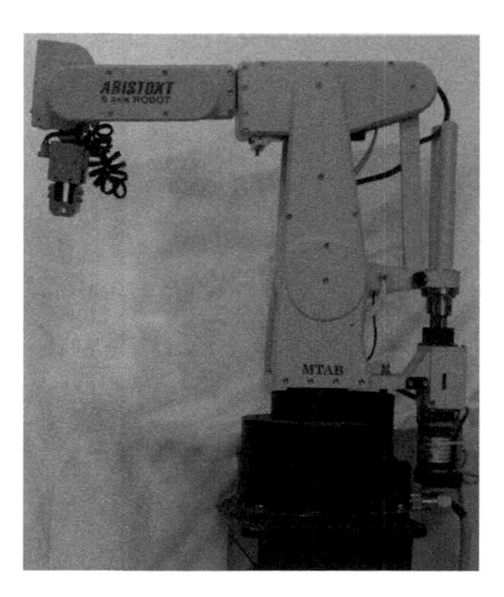

Figure 2.48 A 6-DOF robotic arm

When technology matures, it is expected that concrete 3D printers will be available commercially as off-the-shelf components, just like office printers. Already, many high-level components needed for fabricating printers are commercially available. For example, XY slides for implementing planar movement are available off the shelf. They can be programmed to move in the x and y directions without having to write low-level programs for controlling the motors and other components.

2.9 Summary

The hardware needed for implementing automation is described in this chapter. At the lowest level, basic components such as sensors, actuators, controllers, and communication devices are needed. Sensors collect data that are transmitted to the controllers. Controllers execute programs that analyze the data and take actions. Actions are implemented by sending commands to the actuators in the form of electrical signals. Relays, motors, hydraulic jacks, and solenoid valves are examples of low-level actuators. These components are used to create higher-level devices that are easier to program. High-level devices that have applications in buildings and construction include 3D printers and industrial robots. In complex building and construction systems, several devices are interconnected for coordinated action. Serial and wireless communication are common for integrating sensors, actuators, and controllers into an integrated network.

References

Khoshnevis, B. and Hwang, D. (2006). Contour Crafting. In: Kamrani, A. and Nasr, E.A. (eds) *Rapid Prototyping. Manufacturing Systems Engineering Series*, volume 6, Boston, MA: Springer. https://doi.org/10.1007/0-387-23291-5_9

Soman, Ranjith K., Raphael, Benny, and Varghese, Koshy. (2017). A System Identification Methodology to Monitor Construction Activities Using Structural Responses, *Automation in Construction*, 75, pp. 79–90.

3 Software for automation

3.1 Introduction

Both hardware and software play an important role in automated systems, whether it is for applications inside buildings or for construction activities. Different types of software are needed for implementing automation. At the lowest level, we need software for interfacing with the hardware. At higher levels, we need software for intelligent control and management.

Low-level software is mostly independent of the application. It consists of software for hardware configuration, communication, human-machine interface (HMI), data acquisition, and process control. These are needed by technical personnel who implement automation systems. The methodology for performing these low-level tasks is the same for all applications, whether it is for implementing building automation or for construction automation.

In addition to the low-level automation software, we need software for management tasks and for routine operations. The functions and features of this software are task-specific and involves considerable domain knowledge.

The requirements of software for low-level implementation are first described in this chapter. Then, high-level applications are described separately for building automation and construction automation applications in the next two chapters.

3.2 Low-level software

Primary functions and features of low-level software needed for implementing automation systems include the following:

- Read input produced by sensors
- Produce output for operating actuators
- Support programming the application logic
- Help create and configure the network consisting of controllers, sensors, and actuators
- Enable network communication

3.2.1 Input

As mentioned in the previous chapter, sensors use electrical signals for transmitting the readings. At the lowest level, these signals must be interpreted by the software to take any control action. Modern data acquisition software helps to read data from sensors without

DOI: 10.1201/9781003165620-4

having to write programs. However, popular microcontrollers do not offer this feature. Programs must be written by the software engineer to read and interpret data from the sensors. Examples using some of the popular microcontrollers are provided in this section to understand how this is done.

Sensors typically produce four types of output

a Binary
b Analog
c Digital (serial)
d Resistance

These outputs become input to the controllers.

Binary input

Some sensors need to transmit only a single bit of information. For example, occupancy sensors should tell whether somebody is present in the room. Such sensors produce binary output which is indicated by a high or low voltage (on or off) at its output pin. Most controllers are capable of reading and interpreting high and low voltages on its input channels. The on or off values can be read using the application program interface (API) of the controller. Arduino library has a function called *digitalRead* that returns either HIGH or LOW value, depending on whether there is a voltage at the specified pin. The pin number is passed as the argument to this function.

Reading binary output from sensors such as passive infrared occupancy sensors is simple. Sample code for this is given for illustration:

```
occupancyState = digitalRead(sensorPin);
```

Here, *sensorPin* is an integer whose value is the number of the input pin where the sensor is connected.

Analog input

Sensors that produce analog output require analog-to-digital converters, ADC (See Section 2.7). Many controllers have built-in ADC. If the controllers do not have analog input, external ADC chips and boards could be used for converting analog signals to digital signals. Arduino microcontrollers have several analog input pins. The Arduino library has a function called *analogRead* to read the analog voltage on the specified pin (input channel). The argument to the function is the pin number. Arduino microcontrollers have typically 10-bit ADC. The analog voltage is read as a 10-bit number. With 10 bits we can represent as integer values from 0 to 1023 in binary representation. Therefore, the function *analogRead* returns an integer between 0 and 1023. Since the voltage range for Arduino microcontroller is 0–5 V, the integers 0–1023 are mapped to this range, and the voltage produced by the sensor at the input channel can be calculated. The voltage could be converted into the value of the physical parameter read by the sensor if the calibration curve is known. The user (software developer) should write a program that converts the voltage to the physical parameter and upload it into the controller.

Sample code for reading the voltage from a temperature sensor connected to the pin *sensorPin* and converting it to the actual temperature in deg C is as follows:

```
sensorValue = analogRead(sensorPin);
temperature = 12 + 50 * sensorValue / 1023
```

Here, the voltage 0 corresponds to the minimum temperature 12°C, and the maximum temperature is 62°C (12 + 50).

Digital input

The code to read sensors that produce digital output in the form of serial data are more complex (See Section 2.7.1). Software development libraries contain functions for reading serial data. This is described in Section 3.3.

Reading resistance values

Sensors such as strain gauges produce variable resistance, which have to be converted into voltage or current by using appropriate circuits. In principle, the resistance produced by a sensor can be measured by passing a known current and checking the potential drop across the circuit. But since the change in resistance is usually very small, special circuits are needed to measure it. Wheatstone bridges are typically used (See Section 2.4). This involves the arrangements of resistors that are balanced under normal conditions; there is no output under normal conditions. When the resistance changes due to temperature or strain, the imbalance causes a small voltage or current output. This is amplified and read by the controller. Commercial dataloggers have built-in bridge circuits for reading strain and temperature data. For microcontrollers, these circuits must be added externally. Once, the output produced is compatible with the controller, programming will be the same as that of analog or digital sensors; however, the correct amplification factor should be used in converting the voltage into the sensor readings.

Programming to read input from sensors

For developing simple control applications, a custom program can be written using the API of the controller and uploaded into the controller. Sample code using Arduino microcontrollers is provided in the previous sections. However, for large scale applications, professional software is needed that helps to configure and read input and output from many sensors without having to write programs every time. Historically, this software is known as SCADA. It is the acronym for Supervisory Control and Data Acquisition. It performs functions such as collecting and processing real-time data, logging, visualizing, and analyzing data, etc.

The SCADA software should permit referring to sensors with user-friendly names rather than pin numbers of input channels. It should also allow easily inputting the sensor calibration curve so that voltages can be converted to physical parameters without having to write programs each time. For example, in the BACNET system every sensor is given a unique object identifier, a 32-bit number that represents the type of the object and its instance number. Different types of objects are defined in the protocol such as, Analog Input, Binary Input, Multi-state Input, etc. The object identifiers can be mapped to user friendly names such as "Temperature Sensor in the AHU room". The friendly names could be used in programs, so if the sensor is connected to a different channel, the program need not be changed; only the mapping between friendly names and channel numbers need to be changed in the configuration program. Without this feature, hardware specific information such as "Input channel number 12" have to be specified in the programs.

To replicate these features on low-level microcontrollers, custom programming is needed. First of all, it requires a database (registry) in which the sensor names, locations, and other details are stored. Secondly, the controllers have to be networked and appropriate communication protocols have to be developed so that data can be requested from any controller on the network. Communication modules have to be installed on every controller for listening to requests from the network and responding to them. Since all these involve considerable programming effort, many developers use off-the-shelf commercial SCADA software.

3.2.2 Output

The output produced by a controller is typically of four types

a Binary
b Analog
c Pulse width modulation
d Digital (Serial)

Binary output is supported on all controllers. It produces a high or low voltage at an output channel. Arduino library has a function for this called *digitalWrite*. The pin number and the values HIGH or LOW are specified as arguments to the function. An example of a binary output is a switch that can close or open a circuit. If a small LED light needs to be turned on or off automatically, its circuit can be connected to the binary output channel of a controller. The circuit can be turned on or off by using the function *digitalWrite*.

Analog output is absent in many popular microcontrollers. However, this is supported by traditional PLCs and BACNET DDCs. An alternative for analog output in microcontrollers is *pulse width modulation* (PWM)[1].

PWM works by switching the voltage on and off at a high frequency so that the power delivered to the output channel is reduced. The amount of power delivered is proportional to the duty cycle, that is, the amount of time, the voltage is turned on. The device connected to the output channel sees the average voltage; therefore, it performs like an analog output. For example, the brightness of an LED light can be modified by PWM. Since the frequency of switching is high, flicker is not observed by human eyes and the light appears to be dimmer or brighter depending on the frequency of switching. Arduino library has a function *analogWrite* that uses PWM for producing the effect of variable output. The argument to the function is a number between 0 and 255. The duty cycle is proportional to the specified number.

Another way to send commands to an actuator from a controller is through digital output. Digital output is described in Section 3.3.

3.2.3 Configuring the network

If a single controller is used to connect all the sensors and actuators, configuration is not an issue. However, in a complex automation system, there might be hundreds of sensors and actuators that are physically located in different places. Long wires cause attenuation of signals, and the signals might become difficult to interpret. In any case, the number of input/output channels of a controller is limited. Therefore, several controllers have to be used in the system. Then, the network configuration becomes important. The controllers, sensors, and actuators must be logically and geographically grouped for easy management and maintenance. The network architecture should be carefully designed.

If one sensor is used to control one actuator, it is logical to connect both the sensor and the actuator to the same controller. The control logic can be stored in the same controller and the controller need not communicate with other controllers under normal conditions. However, in practice, control systems must collect data from different sources, and the entire system should be tightly integrated. When the control system network becomes complex, the mode of communication becomes important.

3.2.4 Protocols and standards

A protocol defines a set of rules for facilitating effective communication. It also defines the language of communication. The communication language specifies the format of data transmission. The rules specify, for example, the sequence of steps to be followed for starting a conversation, who has precedence when multiple parties try to communicate at the same time, etc. Communication protocols have been developed for efficient transfer of information between the components of a building automation network such as controllers and workstations. To understand these developments, fundamental concepts related to network communication are briefly explained.

The functions of a communication network are conventionally explained using the open systems interconnection (OSI) model. According to this model, the protocol is divided into seven layers

- Physical
- Data link
- Network
- Transport
- Session
- Presentation
- Application

The physical layer of the communication protocol specifies hardware characteristics. At the physical layer of a wired connection, all the data are transmitted in the binary form as voltage pulses. The pulse width depends on the transmission speed (baud rate). Acceptable speeds are specified in the protocol. As described in Section 2.7.1, two different voltage levels are used to represent one and zero. You need a minimum of 2 wires (positive and negative or ground) to transmit data. You might have additional wires for separating transmitted and received data, as well as for synchronizing communication. Many network topologies are possible. BUS systems are getting popular in building automation. In a BUS system all the devices are connected in parallel to the same communication medium. That is, all the devices use the same set of wires for communication.

The data Link Layer specifies the mechanism for node-to-node data transfer, that is, how two devices that are directly connected to each other communicate. It specifies rules for medium access control (MAC) – how devices in a network gain access to medium. It also controls error checking and frame synchronization (logical link control). Data typically consists of packets of constant size at this layer.

The network layer specifies how data packets are transferred over a network, that is, the mechanism of transferring data between devices that are not directly connected in the network. The scheme for addressing the nodes in the network are defined at this layer.

The transport layer manages the transfer of variable-length data in the form of packets with error control. The session layer defines how a sequence of messages that are part of the same conversation are exchanged between two applications on the network. The presentation layer defines the syntax of data for exchange of information between the applications. Finally, the application layer defines the rules for communication by individual applications.

The OSI model helps to understand conceptually how communication takes place. Each layer in the model packages the data for transmission in a form required by the layer immediately below that. While receiving, each layer accepts data from the layer immediately below and converts it to the required form. This enables the use of different protocols at the lowest layers without changing the implementation at the higher layers. For example, at the physical layer, a USB cable or Ethernet cable might transmit data packets. These are correctly sent and received by lower layers of the protocol stack. The communication module at the application layer need not worry about these details and it simply needs to interpret the received messages.

TCP/IP is a protocol that is widely used in internet communications. It is defined mostly at the transport and network layers, even though all the features of this protocol do not fit into one single layer of the OSI model. Every device on the internet has an IP address, a concept defined by this protocol at the network layer. An IP address is a 32-bit number (in version 4 of the protocol) or a 64-bit number (in version 6) that is used for identifying a node on the network. Whether the addressing uses a 32-bit number or a 64-bit number is defined by the version of the TCP/IP protocol. A 32-bit IP address is usually written as four numbers separated by dots as in 192.188.1.2.

TCP/IP also defines a session-level communication scheme. It contains the concept of a socket that enables reliable exchange of information between two applications. Socket communication has similarities with a telephone call. The person receiving a call should have a unique telephone number. Similarly, an application that wants to accept a connection should create a "server" socket at a specific port. The port is denoted by an integer number that is known to the clients. Different applications running on the same computer are able to use the same network interface (having a unique IP address) for receiving only the data meant for them by using different port numbers. The TCP/IP software separates the data packets received by a computer and delivers them to the specific application that created the socket with the specified port number. The application at the other end creates a "client" socket and connects to the server socket by specifying the IP address and the port number of the server (like dialing a telephone number). Once the connection is accepted, they exchange a series of messages. Whatever data are written by a party at one end of the socket reaches the other party – the TCP/IP software ensures that the receiver reads exactly what the sender has written. Finally, after the conversation is completed, the connection is closed. TCP/IP defines the rules for transferring data packets such that this form of conversation resembling a telephone call is achieved. Internally, lower layers transmit the information in the form of ones and zeros; application programs need not worry about how this is done. A high-level client program written to communicate with sockets looks like this:

Socket object = create socket (IP address, port number)
Write to socket (socket object, message)
Response from server = read from socket (socket object)
Handle response from server
Close socket (socket object)

This high-level program does not contain details about how low-level communication is carried out. This is managed by the underlying protocol stack.

Today TCP/IP protocol stack is implemented on most computers and is part of the operating system. However, low-level controllers may not have built-in support for TCP/IP. In that case, TCP/IP communication should be implemented by including libraries while compiling application programs for these controllers. Arduino has an Ethernet library for internet communication using externally connected Ethernet devices (shields). Using this library, server sockets can be created to listen on specific ports and client sockets can be created to connect to the server sockets.

While TCP/IP software takes care of end-to-end communication between application programs, it is the responsibility of the programs to format the data such that both parties are able to interpret them. Application-level protocols should be developed so that messages that are meaningful to the application can be exchanged. For example, the message "Give me the value of Temperature Sensor 1" should be encoded as a message in a form that a controller is able to understand, so that it can respond appropriately. The application-level protocols are built on top of lower-level protocols like TCP/IP. For instance, data might be sent between two applications through sockets, but the data are formatted and interpreted according to the rules of the application-level protocol.

An example of application-level protocol is HTTP, the familiar one written by Tim Berners-Lee to invent the World Wide Web. Each request sent from a browser to a webserver is formatted in a particular form. For example, when the user clicks on a link to download the page index.html, the following text is sent to the webserver.

GET index.html HTTP/1.1

The webserver reads this message and understands that the file index.html should be sent. It writes the contents of this file to the socket which is read by the browser and displayed on the screen.

Protocols for building automation

In the early days of industrial and building automation, protocols were proprietary and vendor specific. This meant that any expansion or modification to the system can be done only by the same vendor. Interfacing with other systems was nearly impossible. Even accessing data from proprietary systems could be challenging. When open standards were developed, the situation changed. Standards helped improve interoperability and easy communication with disparate systems. Currently popular standards in the building automation domain include,

- BACNet
- LonWorks
- KNX

BACnet stands for Building Automation and Control Network. It is a data communication protocol developed by the American Society of Heating, Refrigerating and Air-Conditioning Engineers (ASHRAE). It is now an ISO standard. BACnet is a true, non-proprietary open protocol communication standard. LonWorks was originally developed by Echelon Corporation and later adopted as an ANSI standard. It is widely used in HVAC control. KNX is an open standard that is supported by many hardware vendors. It was originally called EIB – European Installation Bus. It is widely used in lighting control and home automation. These protocols differ in the cost of implementation, network infrastructure, how data are represented, as well as the communication scheme. However, the actual choice of adopting a particular protocol is often dictated by factors such as the support provided by vendors and contractors for specific applications.

The software for BACnet protocol is available as open source and is widely supported by HVAC system integrators. Therefore, this is taken as an example here to illustrate the features desirable in automation software. The primary requirement for the software is to communicate with all the devices on the network for reading data from sensors and for taking actions by sending commands to actuators. A consistent method of addressing all the devices is essential. A typical BACnet network is shown in Figure 3.1.

Multiple sensors and actuators are connected to a controller. Multiple controllers are connected to a router. Routers, computers, and other IP devices are connected through the Ethernet.

These devices use different physical medium for communication. The router (also known as the network controller) is connected to the internet using a LAN cable, commonly known as CAT5. A small LAN segment might be created using a network switch, by connecting all the Ethernet devices using LAN cables to the same switch. The controllers are connected to the router through twisted pair cables. All the controllers are connected in parallel to the same pins of the controller known as the RS485 interface. The RS485 interface is used to create a "bus" – a common medium through which all the messages are

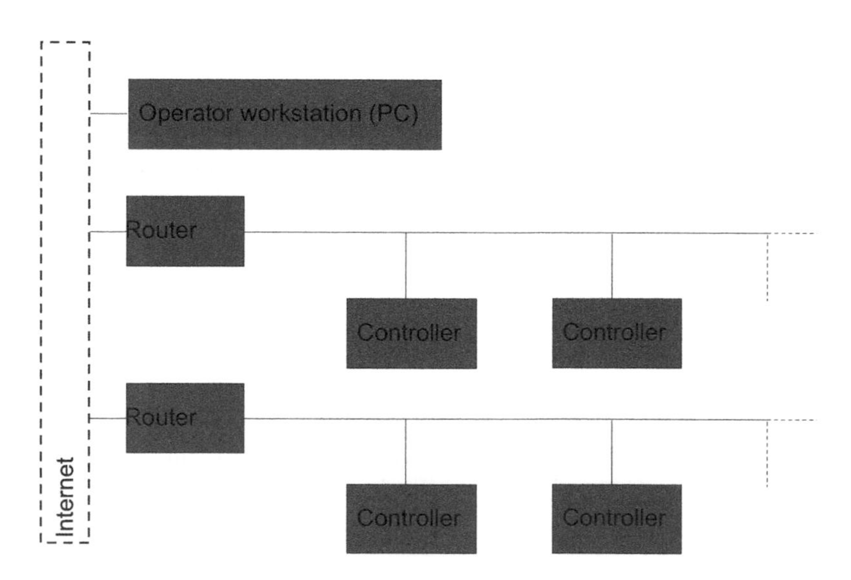

Figure 3.1 A BACnet network

exchanged. The maximum number of controllers that can be connected to a bus is usually around 16.

Every router (network controller) should have an IP address. All the communications from a PC to a router is through the LAN using TCP/IP protocol; actually, using an application layer protocol called BACnet/IP. Individual controllers need not have IP addresses because they do not communicate using TCP/IP protocol. Controllers communicate through the common "bus" using a protocol known as MSTP (Master Slave Token Passing). Routers translate requests from the LAN and transmit them to the MSTP network.

Routers are addressed by their IP address. Since many controllers might be connected to the same router, each controller should have a unique identifier. In BACNET, this identifier is called a Network Identifier or NET ID. In order to communicate to a controller, you need to specify the IP address and the NET ID. Since many sensors (input points) might be connected to the same controller, each input point is given a unique identifier. This identifier is called the BACnet Object Identifier or Instance ID. In order to address a sensor, you need to specify the IP Address, Net ID, and Instance ID.

Apart from routers, several other devices might be present on the local area network of a BACnet system that communicates using the TCP/IP protocol. These devices are also assigned IP addresses that are unique within the local area network. If a computer wants to send a data packet (message) to another computer, it has to specify the IP address of the destination computer. The destination computer uses the IP address of the sender to reply to the message. Similarly, devices such as IP cameras are also provided IP addresses.

3.2.5 Programming methodology

Building automation software controls systems such as lighting, air conditioning, safety, and security. The goal is to achieve optimal performance in terms of energy, occupant comfort, productivity, usability, etc. The control task involves reading environmental and other parameters and then deciding on the control actions based on application logic. The application logic is coded into programs that are executed on the controller. The program might be complex involving thousands of lines of code or in simple cases, a few statements such as the following:

- IF the value of Light Sensor 1 is greater than 500, THEN set Light 1 to OFF
- IF the value of Occupancy Sensor 2 is ON THEN set Light 2 to ON

Most engineers are familiar with programming on a personal computer. A computer program is usually written in a high-level programming language which is then compiled to generate machine code that can be executed by the processor of the computer. The compiler generates instructions that the processor can understand. A program written to run on a controller should also be converted into machine instructions that the controller can execute. The exact procedure for compiling and uploading code into the controller depends on the development environment. Arduino integrated development environment includes a C++ compiler, which generates the machine code for the ATMEGA processor used in Arduino controllers. This code is uploaded into the flash memory of the controller using a USB cable. This code is loaded by the controller at the time of booting (powering up). Raspberry Pi Pico has a similar way of uploading the compiled code into its flash memory using USB cables. However, the common procedure is to upload a Python interpreter, which is automatically started when the controller boots. Then user programs can be written in Python and executed by the Python interpreter.

These procedures become cumbersome when the control system consists of many controllers that run different programs depending on the specific logic related to the sensors and actuators connected to them. Certain BACnet controllers available today use an interpreted language called Plain English for executing application logic. The BACnet software running on a personal computer can be used to upload Plain English programs to any controller on the network without having to individually connect the controllers to the computer with USB cables. There is also support for programming, debugging, and testing distributed programs on the BACnet network. Such software improves productivity while developing large systems.

Typically, control programs run forever. The programs are initialized when the controller powers up; then the user written code is repeatedly called in an infinite loop. Arduino programs usually consist of a *setup* function and a *loop* function. The *setup* function is called only once, after the controller boots. Then the *loop* function is called repeatedly. In addition, interrupt service routines can be set up for automatically executing code when specific events happen. These are functions that are called when the state of input pins change. For example, when the voltage at a pin rises from low to high, a function can be called automatically to attend to that event.

3.3 Digital communication protocols

Basic concepts related to digital communication were introduced in Section 2.7.1. The primary idea in digital communication is to convert data into ones and zeros sent as voltage pulses along a wire. However, several variations are possible; accordingly, several protocols have emerged. Some popular schemes are discussed here.

3.3.1 Serial communication using UART

Serial communication is the most common way of exchanging data between a computer and peripheral devices. Computer mouse and keyboard send data to the computer in this form. Most controllers have at least one set of pins for serial communication. These are usually designated as TX and RX (transmission and receiving). Reading and writing to these pins is done simply by calling predefined functions in the library of the software development kit of the controller. For example, data can be read from the serial port of an Arduino microcontroller using the functions in the serial port object. The following code snippets illustrate its use:

Step 1: Set up the serial communication by specifying the baud rate.

```
Serial.begin(9600);
```

Here the argument to the function *begin* is the baud rate in bits per second.

Step 2: Check if data are available in the serial port for reading. Use the function *available* for this is as follows:

```
If Serial.available()
```

Step 3: Read all the data that is currently available:

```
data = Serial.read();
```

This function returns a byte of data that is read from the serial port.

Similarly for sending data through the serial port, the following functions can be used:

Step 1: Begin the communication:

```
Serial.begin(baudrate)
```

Step 2: Write a text string to the serial port:

```
Serial.write(text)
```

The *Serial* library takes care of converting the data into bits and transmitting them through the wires by applying the correct voltage at the pins at the required speed. All these details are hidden from the programmer.

3.3.2 SPI (serial peripheral interface)

In this protocol, the communication takes place through a four-wire BUS. These wires are used for the following:

- Clock (sending timing pulses for synchronization)
- Output
- Input
- Device selection

These pins are usually denoted using the symbols SCLK (serial clock), MOSI (Master Output Slave Input), MISO (Master Input Slave Output), SS (Slave Select). The SPI protocol follows an architecture which used to be known as the Master-Slave architecture; hence these terms are used in the naming of the pins. The Master is the device that controls the communication by generating the timing pulses. There could be one or more Slave devices. Recently, the terms *master* and *slave* have been replaced by the terms *controller* and *peripheral*, apparently to avoid racial connotations. Since the term controller has been used in a different context in this book, that term is not used in this section to avoid confusion. Moreover, these terms have still not received wider acceptance yet.

In the communication scheme discussed in Section 2.7.1, the pulses are read and interpreted correctly by deciding on the transmission speed a priori. That is, the baud rate is fixed at the beginning of the serial communication. This makes it possible to sample the voltage pulses at regular intervals to convert the electrical signals into bits. In SPI, there is no need to fix the transmission speed a priori. The timing pulses are sent through a separate wire; this is used for synchronizing the communication.

The master device sends the clock pulses at regular intervals in this sequence: the voltage is increased to maximum value, kept constant for a small duration, then it is dropped to zero; then this sequence is repeated. The slave device monitors the voltage on the clock line. When the voltage starts rising, it will start reading the voltage on the data line MOSI and convert it into its bit value. After the master device has sent a message consisting of a certain number of bits to the slave device, it will wait for response from the slave device. The master device will again send clock pulses through the SCLK line. As before, the slave devices will monitor the voltage in the clock line; when the rising of voltage is detected, the required slave device will send the response, one bit at a time during each clock pulse. The data sent back by the slave is through the line MISO. The conversation continues in this manner.

SPI protocol permits communication between one master and multiple slave devices. When multiple slave devices are connected to the same data line, separate wires are used to connect the SS pin of different slave devices to the master device. The master device uses different SS pins. The voltage set by the master device in the SS pin of a slave device indicates that the communication is meant for that particular device. Thus, multiple devices are able to communicate using the same data bus.

Many devices commonly used in automation systems communicate using SPI protocol. For example, SD cards are widely used for storing data. SD card modules used for Arduino microcontrollers use SPI for communication. Arduino controllers have certain designated pins that are used for SPI communication. These pins have to be connected to the corresponding pins of the slave device. Arduino library has simple functions for SPI communication. Sample code snippets illustrating key programming steps are given as follows. Detailed examples are available in the Arduino tutorials.

Code for the master device:

Step 1. Include the header file of the SPI library.

```
#include <SPI.h>
```

Step 2. Initialize the communication using the begin function in the master device

```
SPI.begin ();
```

Step 3. Transfer the data from the master to the slave. Here the data consists of an array of bytes.

```
SPI.transfer (data);
```

Code for the slave device:

Step 1. Include the header file

```
#include <SPI.h>
```

Step 2. Set up the routine to be called when the interrupt (event) is generated, that is, when voltage in the pins change

```
SPI.attachInterrupt();
```

Step 3. Write a function to service the interrupt. In this function, read the data that is received by the slave device.

Other types of microcontrollers have similar low-level software for reading data from devices that communicate using the SPI protocol. These are available in the form of libraries in their respective software development kits. Programs should be written to read and send data to these devices for performing tasks such as acquiring data from sensors or sending commands to devices.

3.3.3 I2C (inter-integrated chip)

I2C protocol was developed to enable communication among multiple devices using the same data bus. Similar to the SPI protocol, there is a clock line and a data line. It also follows a master/slave architecture. However, I2C requires fewer pins than SPI. I2C does not use separate lines for selecting the slave device to communicate to. Instead, every slave device is given a unique 7-bit address. Messages sent by the master device consists of two parts. First, the address of the slave device to whom the message is meant for is sent. Then the actual byte of data are sent. This data packet is read by all the devices connected to the bus. Only the device to whom the data are sent will interpret the data and respond to it. Other devices simply ignore the data packet.

The Arduino Wire library contains functions for communication using I2C protocol. Refer to Arduino documentation for details. Other microcontrollers also have libraries that help in I2C communication.

3.4 Low-level control algorithms

Controllers typically read data from sensors and respond to the external conditions by sending commands to actuators. The step of taking the right decision requires intelligence; in general, it could involve complex logic. Control algorithms compute the set of actions required to achieve a set of objectives. There are many algorithms available depending on the nature of the control task. Some simple options are here.

The typical scenario is shown in Figure 3.2. A physical process that cannot be accurately modeled needs to be controlled. In many cases, there is no model based on physical principles

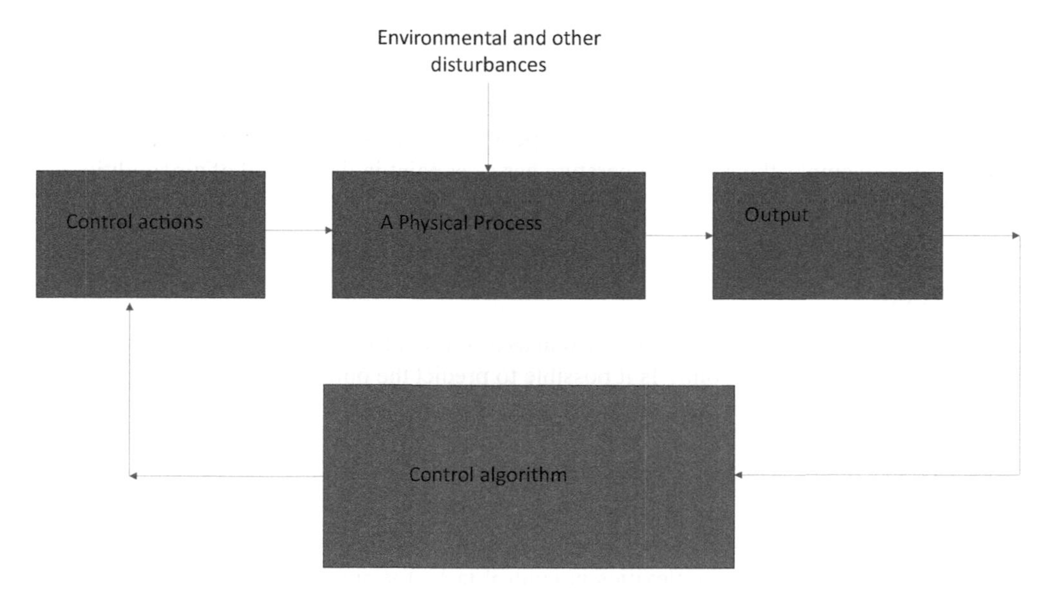

Figure 3.2 Control algorithm

that can accurately predict the output for a given set of inputs. The situation is made more complex by process disturbances. These are factors beyond our control. For example, environmental factors such as heat from the sun and random effects could affect the output from the process. Some of the output variables are measured by sensors. Given this data, the control algorithm should determine what action to be taken to achieve desired results.

There are two strategies for control:

- Open loop control
- Closed-loop feedback control

In open loop control, the action to be taken is decided in a single step. This requires reasonably an accurate model of the process so that the desired state might be reached without an iterative procedure. For example, if you have an exact formula to compute the dim level of a lamp to achieve a required illuminance in the room, the light can be dimmed to the required level in a single action.

In closed-loop feedback control, the desired state is reached in a sequence of steps. A small step is taken, its effects are observed (feedback), and further changes to the action are determined. After each step is taken, the output is analyzed to decide whether to continue with the same action or not. For example, in closed-loop feedback control of a dimmable lamp, the lamp is gradually dimmed until the desired illuminance is achieved in the room. A common difficulty with this process is that the iterative modifications to the input might cause instability (oscillations). An action might cause overshooting the target, which needs to be corrected by reversing the action. This might repeat for long without convergence, resulting in oscillations in the values of the output variables.

From the viewpoint of the control process, the terms *input* and *output* have different meanings from what we are familiar with. What an input is to the controller (hardware) becomes an output to the process; what an output is to the controller is an input to the process. Hence, the terms *input* and *output* should be treated with caution when describing a control process. The output variables of the process (controlled variable) are measured by sensors. For example: temperature in the room is an output of the process, while for a controller, it is an input. The input variables for the control process are actions to be taken, for example, how much the damper should open to bring in fresh air.

In many applications, the desired state is specified by set points on output variables of the process. For example, the room temperature has a set point in the control of air conditioning systems.

Exercise 3.1

Consider the process of studying for an exam as a control task. What are the input and output variables? What is the set point? Is it possible to predict the output? How do you determine what should be the input in order to produce the desired output?

Answer

While students are easily able to appreciate this simple example, the answers are not straightforward. It illustrates the complexities in control tasks for practical applications. Even for tasks that we are familiar with, it is difficult to clearly define the input and output. In this example, there is not much ambiguity about the output; the output is usually the marks scored in the exam. But the input to the process is not easy to define. Students know very well that studying for too long is not always guaranteed to produce better output. In general, we cannot predict what will be the output for a given number of hours of study (input). Some students set targets on the marks they wish to achieve, accepting that they do not want to study all the time. That is, they do not want to maximize the output – they use a set point for the output. The concept of a closed-loop feedback control is relevant to this task. Regular practice tests might be taken after studying certain topics; results of these tests could be used as feedback to determine what changes need to be made in the study pattern for achieving the desired output.

3.4.1 PID algorithm

PID is a standard algorithm for closed-loop feedback control. It stands for Proportional, Integral, Derivative control. It is widely used in many control systems, especially in air conditioning. It is used when the goal is to achieve a set point on certain output parameters such as temperature, and a single variable affects the output. The input variable needs to be changed in a continuous interval for causing the output parameter to move towards the set point. For example, by opening a damper in the range 0–100% you should be able to achieve the set point on temperature in a room.

Exercise 3.2

The valve of a chilled water pipe is currently 20% open. The set point on the temperature of chilled water supplied to an air-handling unit is 7°C. The current value of chilled water supply temperature is 8°C. By how much should the valve be opened to achieve the set point?

Opening the valve causes more chilled water to flow through the pipe and provide more cooling to the air-handling unit. But how much should the valve be open? Suddenly increasing the valve opening to 100% might cause the temperature to suddenly drop below the set point. Opening the valve too slowly might require a long time to reach the set point. The PID algorithm computes the required amount of opening iteratively using feedback as illustrated as follows.

In the PID algorithm, the change to the control variable C is computed as a sum of three terms as follows:

$$C = P + I + D$$

where,

P is the proportional component
I is the integral component
D is the derivative component

The proportional component P

P is proportional to the error (e). The error is the difference between the current value of the output variable and the set point. In example 1, error = 1°C

The P term is written as:

$$P = K_p * e$$

where K_p (proportional gain) is a tuning parameter (which is an input to the algorithm).

In Exercise 3.1, assume a value for Kp = 20 (20% change in the input variable for unit error in temperature). Then,

$$P = Kp\ e = 20$$

Then, the new value of valve opening = 20% + 20% = 40%.

With this new opening, you may not get a temperature of exactly 7°C. It might overshoot the target. Imagine that the temperature becomes 6°C in the next time step. Let us calculate the subsequent valve openings and the resulting temperature using the same formula. Now the error is −1°C. Hence, the new P value will be −20%. According to this calculation, it will revert to the previous opening. Thus, by using the P term alone, we end up with oscillations without convergence.

The integral component I

I is proportional to the integral of the error over time. The integral of the error ($T\Delta_e$) is calculated by summing up the error during each time interval multiplied by the time interval Δt, that is

$$T_e = \sum e \Delta t$$

If Ki is the proportionality constant, I component is written as:

$$I = K_i\ T_e$$

Ki is a tuning parameter like K_p. The integral term considers the long-term effect of applying actions. Even if the error after the action in last time step is negative, if the net error over all the previous time steps is positive, an action is taken proportional to that.

In Exercise 3.2, a PI algorithm is used for control and Ki is chosen to be 10. The valve opening is calculated as follows:

Time Step	*Temperature*	*Valve Opening*	*Error*	*Integral of Error*	*P*	*I*
0	8	20	1		20	
1	6	40	−1	1	−20	−10
2		30				

In the second time step, the integral of error is calculated by summing up the errors in all the previous time steps multiplied by the time intervals. This is multiplied by K_i which is taken as 10. Therefore, the sum of P and I components give −10 as the correction to the valve opening. The opening at the next time step is 40 − 10 = 30. For this opening, the temperature is recorded and the process is repeated.

The derivative component D

D is proportional to the derivative of the error over time. That is,

$$D = K_d \frac{de}{dt}$$

Here, K_d is another tuning parameter. The rate of change of error with respect to time $\frac{de}{dt}$ is used to determine the change in the control variable using the derivative component formula. This component aims to change the output fast if the error increases rapidly.

The PID algorithm is built into many commercially available controllers. For example, many servo controllers use this to attain desired speed or position.

Tuning the parameters

The tuning parameters are usually fixed by trial and error during the implementation of the system. However, several scientific methods have been proposed for tuning the PID parameters to produce robust performance. These can be found in textbooks on control systems.

3.5 Optimization

In general, every control task is an optimization problem. We want to find out the best values of control variables such that some objectives are maximized (or minimized). Optimization methods are discussed in Chapter 7. While the solution to simple optimization problems can be found using exact mathematical methods, for practical engineering problems, optimal solutions must be found by trial and error. This is particularly true in the case of control problems. The relationship between the input and output variables might be highly non-linear and influenced by many effects that are not precisely defined. Accurate models based on physical principles may not be available, and empirical models might have to be developed on the fly using data collected previously. Machine-learning techniques are useful for developing prediction models in such cases. Once a model is developed, best-control actions might be determined by simulating the model. Since an exact global search is time-consuming, approximate solutions are often sought using local search techniques. Local search involves making small changes and moving in the direction that is seen to produce improvements.

3.6 Summary

Software that is needed for implementing automation is discussed in this chapter. Two types of software are needed: low-level automation software and high-level application software. Low-level software is used for interfacing with hardware. This is usually independent of the application domain. This performs generic tasks such as reading sensor data, sending commands to actuators, and supporting program development. These tasks involve communication between sensors, actuators, controllers, and computers. Communication protocols such as UART serial, SPI and I2C are commonly used in automation systems. Such protocols are usually supported by the low-level automation software. Standard control algorithms such as PID are also part of the software development kits of many controllers.

Note

1 Technical terms are in bold italics the first time they are introduced in the book.

4 Application

Building automation

4.1 Introduction

Chapter 1 explained how automation helps in achieving smart buildings. In this chapter, a few examples of building automation are discussed with respect to potential for enhancing building performance. Applying automation in any domain requires detailed technical knowledge that is specific to the domain. In the case of building automation, the performance parameters are mostly related to visual and thermal comfort, and energy consumption. Hence, knowledge of fundamental concepts of building physics related to these performance parameters is essential for using automation for enhancing building performance. For a more detailed treatment of building physics and building performance evaluation, refer to specialist textbooks on these subjects. However, these tend to have a lot more mathematical details than what is needed for the design of automation systems. Hence, this chapter attempts to provide a concise summary of key concepts. The most important performance parameter in building systems is energy consumption. Energy consumption is influenced by many factors related to the detailed mechanical and electrical characteristics of the systems. Without knowing this, it is not easy to appreciate how control actions are able to minimize energy. Hence, this topic is given more emphasis in this chapter.

First, the application areas of lighting and thermal comfort are covered in detail to illustrate how automation helps in enhancing building performance. The following aspects are discussed: what are the performance parameters, what hardware is used and how it improves performance, what software and control algorithms are needed, and how the benefits can be quantified. The final part of the chapter discusses the role of management software for improving building performance.

4.2 Lighting

A building's lighting directly affects the comfort, mood, productivity, health, and safety of its occupants. Electrical lights constitute about 20%–30% of energy consumption of office buildings. Reducing lighting energy is important from the point of view of sustainability. The objective of lighting control is to maintain the required level of lighting in the building, without causing glare and visual discomfort, while minimizing energy consumption. In this section we will examine how these objectives can be achieved.

4.2.1 Basic concepts

Light is a form of energy. It consists of electromagnetic waves that are created by charged particles undergoing acceleration in an electromagnetic field. The waves propagate from the

DOI: 10.1201/9781003165620-5

source and spread out in space until they are absorbed by other charged particles. From the quantum mechanical viewpoint, light consists of particles called photons, which are exchanged between the charges at the source and destination. However, for the discussion about lighting inside buildings, it is enough to consider the classical wave interpretation of light. Wavelength is the most important property of a wave; it is the distance between adjacent peaks in the wave. Frequency of the wave is inversely related to the wavelength by the formula:

$$f = \frac{c}{w} \qquad \text{Eq 4.1}$$

Where f is the frequency, c is the speed of light and w is the wavelength.

Since light consists of electromagnetic energy that flows at a finite speed c, we can compute the amount of energy flowing per unit time, that is, the power in Watts (Joule per second). However, not all electromagnetic energy is visible to us as light. Only electromagnetic waves having a certain wavelength can be received by our eyes and perceived as visible light. This range is called the visible spectrum and consists of wavelengths from 380 to 800 nanometers. Different colors have their ranges of wavelength; violet is from 380 to 450 nanometers on one end, and at the other end of the spectrum, red has wavelengths from 625 to 800 nanometers. Our eyes are more sensitive to certain colors. For example, with the same amount of electromagnetic power, green appears brighter than other colors. Therefore, electromagnetic power (Watts) is not a good quantity to measure the perceived brightness of light. We need another unit that takes into account the sensitivity of the eye to different wavelengths.

Lumen is the unit used to measure the power transmitted by the visible part of light adjusted according to the sensitivity of the eye. This quantity is called the luminous flux. When a lamp converts electrical energy into light, a part of the energy is lost as heat. The remaining energy is converted into electromagnetic waves of different wavelengths. The amount of energy that is transmitted per unit time in the visible range is measured by the quantity, luminous flux, and expressed in units of lumen. To get an idea of the scale of a lumen, consider a hypothetical lamp that converts 1 watt of electrical power entirely into light of wavelength 555 nanometers. This lamp produces 683 lumens. This is the maximum light that can be produced by any hypothetical lamp – it has a maximum luminous efficiency of 683 lumen per watt. In practice, luminous efficiency of actual lamps is much lower. An efficient LED lamp produces about 100 lumens per watt. Old incandescent lamps have an efficiency of about 20 lumens per watt. Today, high-efficient fluorescent lamps produce more than 100 lumen per watt of electricity consumed. For example, a compact fluorescent lamp (CFL) of 18 Watts is capable of producing 1,800 lumens.

Light that is emitted by a source goes in all directions. The power is distributed over a range of solid angles. We are interested in finding out the power that is emitted in a particular direction. The unit *candela* is used to measure this. It measures a quantity known as luminous intensity, which is the visible power transmitted per unit solid angle. That is, candela = lumen/steradian.

When light hits a surface, part of it is reflected, and part of it is absorbed by the surface. The light reflected by the surface reaches our eyes. If more power of light reaches our eyes from the surface, it appears to be brighter. Our perception of the brightness of a surface depends on the light power that hits the surface per unit area (this in turn determines the amount of light reflected by the surface into our eyes). This quantity is called illuminance and is measured in units of lux. 1 lux = 1 lumen per square meter. Note: foot-candle is another unit for illuminance (1 fc = 10.764 lux).

Imagine that a bulb is at the center of a sphere of radius 1 m. If the bulb gives out 2,800 lumen equally in all directions, the interior surface of the sphere will receive ($2800/4\pi$) lumens per square meter, where 4π is the surface area of the sphere of radius 1 m. Therefore, the surface will have an illuminance of 223 lux.

Exercise 4.1

This exercise helps in understanding the concepts of lumen, lux, and daylight transmission. An office space has a lighting layout in the form of a grid of spacing 2 m. Each luminaire consists of two 28 W bulbs producing an output of 110 lumen per watt. The ceiling has reflectivity of 80% and is at a height of 2.2 m from the floor. Natural daylight coming through windows consisting of a minimum of 100 lux is available throughout the room. Calculate the illuminance at the work plane at a point directly under a luminaire.

Assume that the work plane is at a height of 0.6 m from the floor, and the light from the bulb is distributed uniformly in all directions. This is a simplifying assumption for getting an approximate estimate of the brightness level. For accurate calculations, a lighting simulation software can be used.

Solution

The cross section of the room is shown in Figure 4.1. Only three luminaires are shown: A, B, and C. The work plane is at D. To calculate the lighting power at D from luminaire A, assume that the light is uniformly distributed along a hypothetical sphere of radius 1.6 m. Thus, the direct contribution from luminaire A is

$$L1 = \frac{2 \times 28 \times 110}{4\pi\ 1.6^2} = 191.5$$

Since half the sphere is above the ceiling, the light which is going in this direction will be reflected down. If there are no losses and the ceiling has 100% reflectivity, we would expect the contribution from A will be doubled. However, since the ceiling has only 80% reflectivity, we can approximately calculate the reflected component as L2 = 0.8 * 191.5 = 153.2.

The contribution from the adjacent luminaire B is calculated as

$$L3 = \frac{2 \times 28 \times 110}{4\pi\ 2.56^2} \times \left(\frac{1.6}{2.56}\right) = 43.8$$

Note that the expression involves multiplication by cosine of the angle to the normal because radiant power from luminaire B is inclined to the surface. Since there are four luminaires at a distance 2m from D in the two horizontal directions, the contribution from L3 should be multiplied by 4. The reflected components from these luminaires are ignored because of the large angles. The contribution from remaining luminaires is likely to be small because of the large distance from D. Therefore, the total illuminance at D from the luminaires can be approximated to be 191.5 + 153.2 + 4 * 43.8 = 520 lux. (Note that 56-Watt lamps at 2m spacing produces around 500 lux, which is the commonly recommended illuminance in office spaces). Adding the daylight component, the total illuminance at D = 620 lux.

Recommended illuminance

Codes specify the minimum amount of illuminance for different types of buildings. For classrooms, it is around 300 lux, and for offices it is about 500 lux. For special tasks such as watch repair that require concentration, you need much higher illuminance. For general visibility in corridors and circulation areas, the required illuminance is around 100 lux. The maximum illuminance inside rooms is limited to around 2000 lux, because high illuminance might cause problems due to reflections and glare.

4.2.2 Lighting system design

The lighting system is designed after determining the visual needs of the space. The type of lighting to be installed and their lighting power should be determined. There are mainly four types of lighting in buildings:

- Ambient lighting – for circulation and general lighting
- Task lighting – used where clearly defined lighting levels are required for detailed work
- Accent lighting – for architectural purposes to add emphasis or focus to a space or highlight a display
- Emergency or egress lighting – to provide a pathway for emergency escape

Ambient lighting is usually provided in the form of electrical lamps installed on ceilings. Ambient lighting tends to consume a significant percentage of energy used for electrical lights in buildings, especially in large commercial and office buildings. The energy for ambient lighting can be reduced by using these rules:

a Lower ambient lighting levels can be used if occupants use task lighting to do most of their work
b Higher ambient lighting levels are provided in densely occupied work areas

Task lighting is provided to supplement ambient lighting to cater for increased task specific lighting requirements. It is provided by means of table lamps and lighting integrated into the furniture.

Exercise 4.2

A numerical exercise is taken to understand the energy saving potential of task lighting. Two options for lighting in an office space of 9 m × 9 m are given in the following. Discuss which is better.

Option 1: Ceiling mounted lighting fixtures along a grid of spacing of 2 m. Each fixture consists of two 28-Watt fluorescent lamps. This layout produces an illuminance of 600 lux on the work plane.

Option 2: Ceiling mounted lighting fixtures along a grid of spacing 2 m. Each fixture consists of one 28-Watt fluorescent lamp. This layout produces an illuminance of 300 lux on

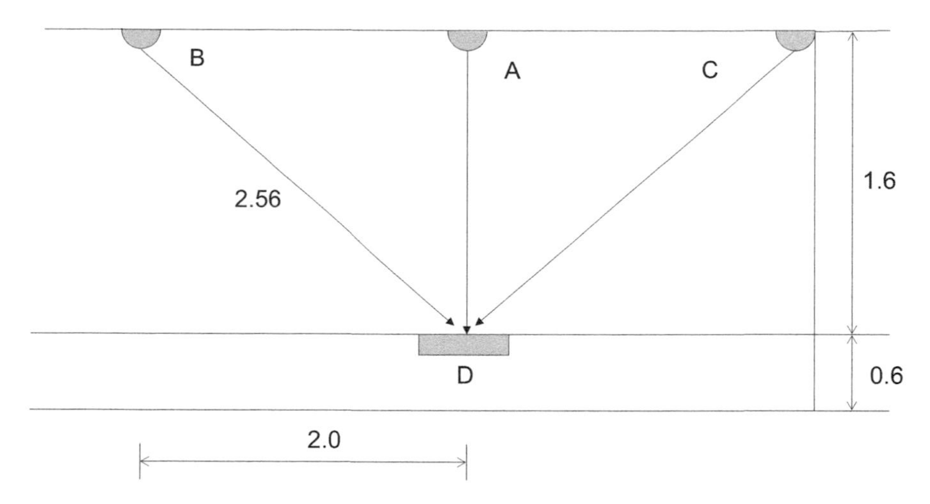

Figure 4.1 Calculating illuminance

the work plane. In addition, there are 6-Watt task lights on every desk. There are 9 desks in total. Task lights produce 300 lux on the work plane.

Solution

If we leave one meter space on all the four sides of the room, there are five rows of lights at a spacing of 2 m. The total number of lights on the ceiling is 5 × 5 = 25. The total lighting power consumption for option 1 is: 2 × 28 × 25 = 1,400 Watts.

In option 2, the power of ceiling mounted lights is 1 × 28 × 25 = 700 Watts. The power consumed by task lights is 6 × 9 = 54 Watts. The total power is therefore 754 Watts.

Option 2 is better than Option 1 in terms of power consumption. In both cases, the illuminance at the work plane is 600 lux. Task lights are more energy efficient because the lights are closer to the work plane and they illuminate a small area on the desk instead of lighting the entire volume of the room. In this example, 6-W task lighting is able to produce the same illuminance as 28-W ceiling lights.

4.2.3 Basic lighting control strategies

The energy consumption of electrical lights can be optimized using appropriate control strategies. Visual comfort can also be improved. The common way of controlling the lights is by providing manual switches that can be turned on or off. An improvement to this is using timer-based controls. Timer-based switches are designed to be turned on manually and turned off automatically after a specified time. Both mechanical and electronic timers are available. There are also models that can be programmed to turn on and off at specified times. These are used in places that are briefly occupied such as toilets and plant rooms, where people enter for a short duration. They might forget to turn off the lights when they leave, and energy is wasted.

Manual dimmer switches could be used to reduce the brightness when required. Dimmable ballasts are needed in fluorescent lighting to reduce the brightness. Dimmer switches are usually provided in meeting rooms, lecture theatres, and auditoriums to reduce the lighting level during presentations. Apart from improving visual comfort, it could also reduce power consumption by reducing brightness when full brightness is not needed.

Simple techniques can be used to reduce lighting energy consumption when full lighting power is not needed in situations such as lunch breaks and after office hours. Reduced lighting is acceptable after office hours for security personnel or cleaning crew. Some of the lamps in a multi-lamp fixture can be turned off by separating the electrical circuits and providing separate switches. In large halls, alternate lamps in a lighting grid can be connected to separate circuits and could be turned off together. This might affect the uniformity of lighting in the room, but it could be adopted when lighting is needed only for general circulation. Lamps near the window should have separate switches so that they can be turned off when adequate daylight is available.

4.2.4 Computerized control of lights: automated lights

There are two approaches to automated control of lights.

a Using a Building Management System (BMS)
b Using sensors

Lighting can be programmed to start and stop at times specified in the BMS. This is referred to as lighting sweep strategy. This is most appropriate in offices and commercial buildings which have well-defined working hours. However, it is necessary to provide a manual override feature so that users who want to stay longer in the building should not be put to inconvenience. Legitimate users could be given access to the BMS so that they can set the time themselves. Otherwise, switches could be provided near their workplaces so that they can request an extension of time. More fancy methods include voice-activated switches (users shout or scream when lights get turned off!), cameras with motion detection (users wave at the camera to attract its attention), and occupancy sensors to automatically detect whether people are in the room.

Sensors are increasingly being used to control lights. Two approaches are common:

a Use of occupancy sensors
b Daylight harvesting using light sensors

Occupancy sensors

Passive infrared (PIR) motion sensors and ultrasonic can be used to detect occupancy. PIR sensors detect changes in the infrared patterns in the room to infer the presence of occupants. Lights might be turned off or dimmed when occupants are not present. However, the following issues should be adequately addressed in the implementation:

- There might be blind spots in the room that are outside the range covered by the sensor or blocked by furniture and other objects. To prevent this, the locations of sensors must be carefully selected. The system should be thoroughly tested during a long period of time before occupants move in. All possible blind spots and other obstructions should be detected.
- People may remain still in the room, and the sensor might not detect any motion. To avoid this problem, a time delay should be set. Lights should be switched off only after a certain time during which no motion is detected

Daylight harvesting

In the concept of daylight harvesting, natural daylight is used to reduce the energy consumption of artificial lights. Lights are dimmed or switched off when sufficient daylight is available. The amount of daylight available is detected using light sensors (photo sensors). This should be used in conjunction with good daylighting design so that maximum daylight is allowed into the room.

How much light should be supplied by electrical lights when some daylight is available?

Let the target illuminance on the work plane be T. Note that T depends on the intended use of the space. Let D be the illuminance on the work plane due to daylight coming into the room. In practice, the daylight component of illuminance cannot be measured directly unless all the lights are switched off. If D is greater than T, there is no need for electrical lighting. Lights can be switched off. If D is less than T, the daylight must be supplemented by electrical lighting. The electrical lights should produce an illuminance of (T − D), which is the deficit in the lighting level. If the electrical lights are designed to produce an illuminance of L, the percentage of brightness of lights is calculated as 100* (T − D)/L. Dimmers are provided control signals to reduce to brightness to this level.

Different types of dimmers are commercially available. There are electronic dimmable ballasts that take analog command input in the range 1–10 VDC. Based on the input voltage, the dim levels are calculated by linear interpolation and the lights are dimmed accordingly. Another popular system is Digital Addressable Lighting Interface (DALI). DALI is a protocol for communication between electronic ballasts and sensors, based on digital communication rather than analog interfaces.

An energy efficient means of dimming lights is through ***Pulse Width Modulation*** (PWM). In simple terms, it involves reducing the time during which power is available to the circuit by turning on or turning off the power supply periodically at a fast rate. Therefore, power is provided to the device only during a reduced duration of time, called the duty cycle (width of the electrical pulse). Thus, the energy consumption is reduced. Since the average power provided to the lamps is less, it appears to be less bright. Since the switching happens faster than our eyes can detect, we do not notice any flicker.

Exercise 4.3

The electrical lighting in an office space is designed for 600 lux. The required illuminance in the office is 500 lux. At 11 am in the morning the illuminance on the work plane due to natural daylight is 250 lux. To what brightness level (in percentage) can the lights be dimmed?

Answer

The deficit in lighting is (500 − 250) = 250 lux. The lights can be dimmed to (250/600) = 41.67% brightness.

Power consumption of dimmable lights

It is necessary to quantify the energy savings to justify the cost of dimming controls. The power consumption of electrical lighting consists of the power consumed by the lamp and ballast losses. The power consumed by the lamp varies with the percentage of dimming. An ideal dimmer has a linear relationship between power and the percentage of dimming. This is possible through pulse width modulation. Some commercially available dimmers cannot

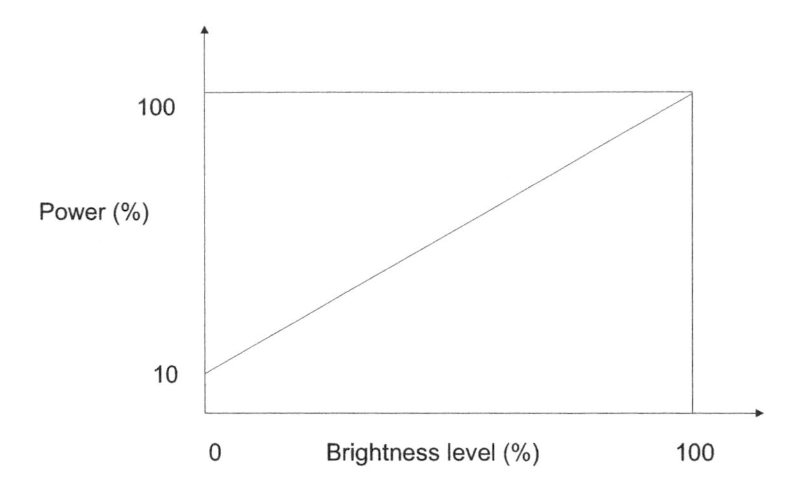

Figure 4.2 A linear dimming curve

reduce the brightness below a certain percentage. For example, the dimming curve of a dimmable ballast is shown in Figure 4.2. The minimum power consumption is 10% at 0% brightness because of the power consumed in the circuits. The power consumption increases linearly from 10% to 100% for full brightness.

Exercise 4.4

The electrical lighting in an office space is designed for 600 lux. The required illuminance in the office is 500 lux. At 11 am in the morning the illuminance on the work plane due to natural daylight is 250 lux. A lamp of 56 W is dimmed to obtain target illuminance. What is the power consumption of this lamp? Use the dimmer curve in Figure 4.2. Assume there are constant ballast losses of 4 W (irrespective of the dimming level). What is the percentage savings in energy due to dimming?

From Exercise 4.3, the percentage brightness required is 41.67%. Interpolating between 10% and 100%, the power consumption corresponding to this brightness level is $(10 + 90 * 0.4167) = 47.5\%$ Therefore, the power consumed by the lamp including the ballast losses is $56 \times 0.475 + 4 = 30.6$ W. If the light is not dimmed, it would consume $(56 + 4) = 60$ Watts. Therefore, we achieve an energy savings of $(60 - 30.6)/60 = 49\%$

Issues in daylight harvesting

Ideally, the lights should be dimmed to the right level according to the availability of daylight at the work plane. However, this is difficult to measure. Light sensors are typically installed on the ceiling. It is not practical to put sensors on the desks because they will be obstructed by papers and other objects that are needed for work. Ceiling mounted sensors cannot directly measure the illuminance at the work plane. Sensors have a wide coverage and measure the average illuminance over the range of angles covered by the sensor. It cannot measure the illuminance at a single point on the desk. Also, it is impractical to measure the illuminance due to the daylight component alone. The light sensor measures the total illuminance, which is the sum of daylight and electrical light components.

Another problem is that light sensors might pick up local reflections at certain times of the day, and they read values higher than the actual illuminance on the work plane. If a shiny object reflects light directly to the sensor the reading will be very different from that at the work plane.

A potential solution to the problem of reading the illuminance at the work plane is given in the following: At night when no daylight is available, lights are turned on at full power, and the readings of the sensor on the ceiling are recorded. Simultaneously, a lux meter is used to measure the illuminance at the work plane. Using a few points taken under different dimming levels, the correlation between the work plane illuminance and the sensor measurement can be calculated. Knowing the current dimming level of the ballast, the component of illuminance due to electrical light can be calculated using the correlation coefficient. Similarly, measurements are taken during daytime under different conditions, and correlation between readings of lux meters at the work plane and those of the sensors on the ceiling is calculated. Knowing this relationship, the daylight component of the sensor reading can be calculated after subtracting the component of the electrical lights. This helps to predict what will be the illuminance if the lights are completely switched off. This can be used for open-loop control. Commercial dimming control systems use closed-loop feedback control. They adjust the dim level iteratively in a feedback loop to achieve the target. However, even these need the correlation between readings of the sensors on the ceiling and the work plane illuminance so that the target illuminance at the work plane is achieved.

The problem of local reflections might be solved by using multiple sensors to control the same set of lamps. However, care should be taken not to take the average of multiple sensors. If local reflections cause large increases in sensor readings, the average is likely to be high, and the resulting decision will be to turn off the lights. Instead, the minimum value recorded by all the sensors should be taken to be on the conservative side. This way, the effects of local reflections on individual sensors can be eliminated from the control actions.

Dimming control has the potential to reduce energy consumption. However, it should not be at the cost of occupant comfort. Rapid changes in lighting levels cause discomfort to users. Imagine that the clouds cover the sun suddenly, and daylight levels drop significantly. The lighting control system might take some time to respond to this event, and light levels remain low for a short period. If the brightness of artificial lights is increased suddenly, it will cause visual discomfort. The dimming system should gradually turn on the lights so that the transition is smooth. When the clouds disappear, the lights should be switched off gradually. If the changes are sudden and this repeats frequently, users are likely to be distracted and irritated. These type of conditions influences user acceptance of automated controls. It is a challenge to ensure that the lighting system responds fast, but the brightness levels are adjusted gradually.

4.2.5 Control of daylighting features

Daylighting features can be passive or active. The term "passive" denotes features that are static and do not respond to the changes in the environment. Active features make use of automation and control. Both passive and active features could be used to create high performing smart buildings. The focus of this section is actively controlled daylighting features. However, it is emphasized that the active features work best when used in conjunction with passive features.

Passive daylighting features include skylights, windows, shading, light shelves, light pipes, and mirror ducts. How these features work together with automated systems to create a smart building is discussed here.

If skylights, windows, and shadings are properly designed, they could improve daylighting in the interior without causing glare and bringing too much solar radiation into the room. Improving daylighting and reducing the heat transmitted into buildings are often conflicting objectives. Solutions should make reasonable trade-offs among these objectives. Adaptively controlled devices have better capabilities to balance the different needs of occupants. Many passive daylighting features can be actively controlled to improve their performance. Some of the possibilities are discussed here.

Light shelves

A light shelf is a horizontal or inclined projection with a high reflectivity meant to increase the depth of daylight penetration into a room (Warrier and Raphael, 2017). It operates by reflecting sunlight off to the ceiling from where it is further reflected to the work plane (Figure 4.3).

Most light shelves are static – they do not move. With a fixed geometry of light shelves, it is not possible to respond to the changes in the position of the sun. Energy savings and general performance can be improved by adapting the geometry to environmental conditions. In the evenings when the sun is at a low altitude, light shelves are not capable of blocking direct sunlight. It could cause glare inside the room. Window blinds are usually used to prevent this. Another possibility is to rotate the light shelf to block direct sunlight, and thus improve visual comfort (Raphael, 2011). The efficiency of light transmission can also be improved by making the shelf rotatable. If the light shelves remain horizontal, they are able to increase daylight penetration only when the sun is at an appropriate low angle. When the sun is vertically above, the reflected light goes back to the sky. By making the external light shelf rotatable, its angle can be adjusted to the position of the sun to produce deeper daylight penetration.

A simple system with one degree of freedom helps to improve the performance of a light shelf significantly. A stepper motor is sufficient to rotate the light shelf to the required angle. However, the control algorithm must be robust enough to account for a wide range of conditions: cloudy conditions, different hours of the day, days of the year, user acceptability of frequent movements of the shelf, etc.

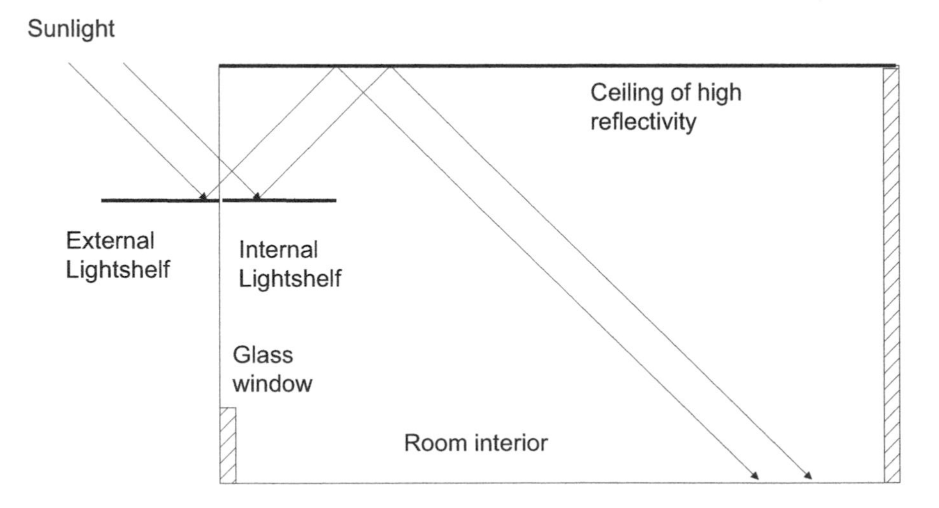

Figure 4.3 A light shelf

Solar trackers have been used in photovoltaic systems for maximizing the power production. These solutions are not appropriate for light shelves because maximizing the sunlight is not always required. Other parameters such as glare and uniformity of lighting are more important. Optimal control should consider all these parameters.

Light pipes

Light pipes are devices that collect sunlight from the roof and transmit it to the interior of the building through pipes having a reflective internal surface. It has a collector at the top designed to maximize the amount of solar light that is transmitted through the pipe. Lenses have been used to focus light into the light pipes. Sometimes, there is a dome with a rotating mirror which is oriented according to the position of the sun. The mirror is rotated such that maximum sunlight is reflected into the interior of the pipe. These are already commercially available as finished products. The pipe is made of steel or other construction materials. The internal coating of the pipe should be highly reflective because light undergoes multiple reflections before it is transmitted into the room. At each reflection, a certain percentage of light is lost through absorption by the material. Therefore, the light level drops fast. Coatings with more than 99% reflectivity are available; these are required to ensure reasonable light throughput of the pipe. At the bottom of the pipe, there is a diffuser to ensure that the light that is coming out is uniformly distributed. Otherwise, there will be problems due to glare and visual discomfort.

Straight light pipes produce too much light and heat in tropical regions. Bent pipes increase the number of reflections of light rays and tend to reduce the illuminance significantly. Bent pipes are needed when openings are not practically feasible on the upper floors at the locations where diffusers need to be placed inside rooms of lower floors.

Automation of light pipes involves the use of sensors to ensure that maximum light is collected and transmitted into the room. Sometimes, it may also be needed to block excess light coming out of the pipe. The position of the sun might be tracked to rotate the mirror and maximize the light transmission. This might be used in conjunction with light dimmers so that energy is saved when adequate light is available. The lighting fixture might be integrated with the light pipe diffuser such that occupants are not aware whether the light is coming from electrical bulbs or from the sun; this improves the user experience. The light should be dimmed such that a constant illuminance is available all the time. Smart control is needed for this.

Mirror ducts

A mirror duct is also a light-redirecting device like a light pipe. Mirror ducts collect sunlight from the edge of the building and transmit them through horizontal ducts having a reflective interior surface. The light is supplied to the room through diffusers, which ensure a uniform distribution of illuminance. The diffusers could be designed to look like electric lighting fixtures or could be integrated with lighting fixtures. The electric lights can be dimmed according to available light from the diffusers. Users may not notice the difference between electric light and daylight, and this helps to improve user experience.

The daylight collectors of mirror ducts could be adaptively controlled to maximize the transmission into the room. Mirrors could be used to redirect more light into the duct, and these could be adjusted according to the position of the sun.

The benefits of mirror ducts should be carefully evaluated against the cost. Due to multiple reflections, the output from mirror ducts tends to be low, especially in the case of lengthy pipes. Active control increases the cost further. Savings in the cost of electricity may not be adequate to justify the cost of installing mirror ducts. Other considerations such as sustainability might be more valuable for certain stakeholders, and this would influence the choice even when cost is not attractive.

Window blinds

Commonly used window blinds are shading devices meant to reduce the amount of sunlight transmitted into the room. If the blinds are manually operated, the responsibility of opening and closing them rests with users. When it is not too bright outside, they open the blinds to bring in more daylight and to ensure good visibility outside. They close it when high brightness is not visually comfortable. But people tend to leave the blinds closed even when the brightness level is not too high, because of the manual effort needed to open the blinds. This results in artificial lights remaining on unnecessarily, wasting energy. Automated blinds help to eliminate this problem.

There are many variations possible in the designs of motorized blinds: roller blinds, vertical and horizontal slats, etc. These are commercially available and can be installed easily. These can also be integrated with lighting control systems using protocols such as EIB/KNX. If they are linked to the BMS, their positions can be monitored in real time through the BMS. The data from BMS can be analyzed to quantify the energy savings and other parameters.

Control of shades

Glass facades are aesthetically attractive, and many architects prefer that. It also improves daylight inside buildings. However, glass tends to transmit too much heat into buildings. Glass with special coatings is used to reflect excess heat and reduce the heat ingress. An example can be seen in Figure 4.4. Buildings with glass façades look good, but several

Figure 4.4 Reflective glass on a façade

Figure 4.5 Shading panels on a roof

problems have been reported. In one case, a building had a curved surface, and the reflected heat from the façade set a car ablaze that happened to be at the focal point of the reflected rays. Environmentalists object to the use of reflective glass since birds get disoriented and confused by the reflections. Reflective coatings also affect the light transmission. If the coatings are spectrally sensitive, it could selectively transmit visible light inside without too much heat. However, in general, reducing heat also reduces the visible light transmittance.

The best option to reduce heat inside buildings is by using opaque shades. Like light shelves, they can also be actively controlled. When excess radiant heat is not present, the shades could be put away by moving or rotating it so that more light can be transmitted. When heat needs to be reduced, shades could be made to cover the glass and block the radiation. This strategy could be used on skylights and openings in ceilings. An example from an airport building is shown in Figure 4.5. A large array of panels is provided on the roof to control the amount of heat and light according to requirements at the specific time of the day.

4.2.6 Quantifying the benefits of automation of daylighting features

Energy savings and visual comfort are the main parameters that could be used to quantify the benefits of daylighting features. Energy savings result from reduced power consumption of artificial lights when daylight is available. Energy savings are computed like this:

- Calculate the annual energy consumption when the lights are on all the time at full power, that is, without using any daylighting features.
- Calculate the annual energy consumption assuming that the lights are dimmed or switched off when adequate daylight is available
- The difference between the calculations gives the energy savings through the use of daylighting feature

Exercise 4.5

a Calculate the annual energy consumption of electrical lights in Option 1 of Exercise 4.1. The office is operational from 8 am to 6 pm on all weekdays.

b Assume that one row of lights near the window can be switched off for two hours every day since adequate daylight is available in this area. This could be done using the lighting sweep strategy programmed in the BMS. Calculate the annual energy consumption of electrical lights and the percentage savings through the lighting sweep strategy.

c Assume that one row of lights near the window can be switched off from 12 noon to 2 pm every day and dimmed to 50% from 2 pm to 4 pm. The dimming is done automatically using light sensors, but since it is difficult to predict the actual level of dimming, the 50% assumption is made. Calculate the annual energy consumption of electrical lights and the percentage savings in lighting energy.

Answer

a The annual energy consumption is calculated by multiplying the total power by the number of hours. Since the power consumption of electrical lights is 1.4 kW in Exercise 4.1, the annual energy consumption is: 1.4 × 10 hours per day × 5 days per week × 52 weeks in a year = 3,640 kWh.

b Reducing the number of hours of operation of one row of electrical lights, the annual energy consumption is 0.056 × (5 x 8 + 20 × 10) × 5 × 52 = 3,494 kWh. The savings in energy through light sweep strategy is 4%.

c Assuming the linear dimming curve given in Figure 4.2, the power consumption at 50% dimming is 55%. The annual energy consumption is calculated as 0.056 × (5 × 6 + 0.55 × 5 × 2 + 20 × 10) × 5 × 52 = 3,429 kWh. The energy savings is 5.8%

Lighting simulation

The amount of daylight available in a building can be computed using lighting simulation software such as Radiance (Ward and Shakespeare, 1998). The illuminance at any point within the building (or exterior) at any time of the day is computed by tracing the rays of light from the source to the point of interest. The simulation software uses the sun path for the location and a model of the sky to calculate the amount of light that is transmitted into the building. Users should define the 3D geometry of the building as well as material properties such as reflectivity of surfaces and light transmission coefficients of glass.

Knowing the amount of daylight available, the percentage of dimming of lamps and the power consumption can be calculated. This permits calculation of the energy savings through the use of daylighting features.

4.3 Thermal comfort

Thermal comfort has direct influence on productivity and efficiency of occupants in a building. Uncomfortable work conditions cause absenteeism, sickness, and disinterest in work. Therefore, most commercial and office buildings are designed to provide a comfortable environment for occupants, usually through air conditioning. Air conditioning systems consume about 40%–50% of energy in commercial and office buildings. Therefore, the potential for energy savings through efficient control of air conditioning systems is huge. Energy savings arise from more efficient operation of existing equipment, reduced hours of operation, efficient management, and the use of novel energy efficient equipment and strategies. This section discusses ways to increase thermal comfort and energy efficiency through automation. First, basic concepts related to air conditioning are reviewed for a better understanding of the potential for automation.

4.3.1 Review of concepts related to air conditioning

There are many types of air conditioning systems, from split units to centralized systems. The common principle used is evaporative cooling, making use of the fact that liquids absorb heat while undergoing phase change to become gas. The amount of heat absorbed in this process is called the latent heat of vaporization. If the liquid is forced to evaporate by applying a suction pressure, it boils at a lower boiling point, and in the process, absorbs latent heat from the surrounding. The temperature of the surrounding material drops because of this absorption of heat. The cooling thus produced can be transferred to an appropriate material and transported to locations where cooling is required.

Air conditioning systems use refrigerants that have a low boiling point so that they can easily be converted into gas through forced evaporation at low pressure. A suction pump creates low pressure to maximize the amount of liquid that undergoes phase change. When the refrigerant evaporates, the temperature drops. The mixture of refrigerant in liquid and gaseous states is passed around coils, and the large area of the coils is used to absorb the cooling. Then the refrigerant is converted back to a liquid state to repeat the cycle. The refrigerant is forced to undergo phase change from gas to liquid by applying a high pressure, the reverse process of forced evaporation. In this process, it releases the latent heat that was absorbed during evaporation. The released heat is absorbed by an appropriate medium and dissipated into the atmosphere.

In commonly used split units, air is directly used for absorbing the cooling from the refrigerant. In split unit air conditioning, there are two components: the indoor unit and the outdoor unit. Inside the indoor unit, air is blown over the coils circulating the cold refrigerant. This cold air is then pumped into the room. The refrigerant in the gaseous state is converted back to liquid state through a condenser that is part of the outdoor unit. The heat released by the condenser is released outside by blowing air. Thus, the split unit system pumps heat from inside the building to the outside.

In centralized air conditioning systems, chillers are used for producing cooling. The standard process of cooling, called the *vapor compression refrigeration cycle* is illustrated in Figure 4.6. In this cycle, there is evaporation on one side and condensation on the other side. Evaporation produces cooling; condensation produces heat. The chiller contains a pump for creating low pressure to force the phase change of the refrigerant in the evaporator. The cooling thus produced is typically absorbed by circulating water in coils within a container (shell) carrying cold refrigerant. Chilled water pumps are used to circulate the cold water to air-handling units (AHU) where air is cooled using this chilled water. To convert the gaseous refrigerant into a liquid state, high pressure is applied inside the condenser by the chiller pump (note that this is not the same as the chilled water pump). Usually, the same chiller pump is used to produce both the low pressure and the high pressure on two sides of the pipe carrying the refrigerant. This is done using an expansion valve, a tiny opening that allows gas to expand suddenly to create low pressure. Heat produced during the forced condensation of the refrigerant is absorbed by water that is circulated in coils. This hot water is pumped by the condenser water pump to a cooling tower where the heat is

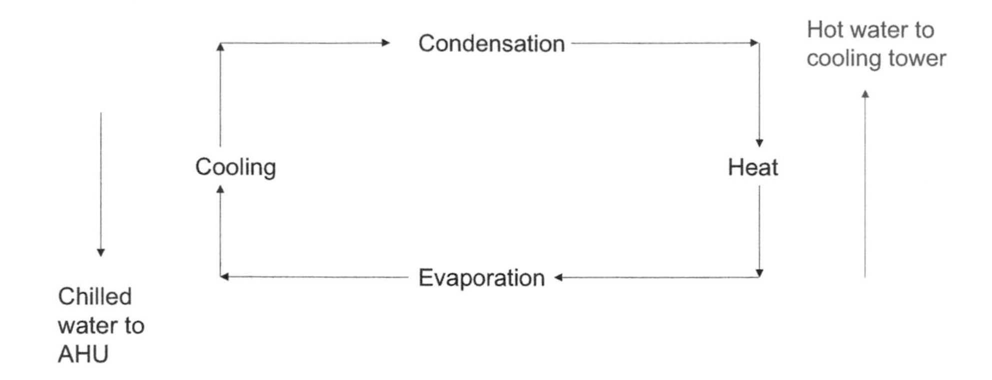

Figure 4.6 Vapor compression refrigeration cycle

rejected. The common method is evaporative cooling by sprinkling hot water and blowing air over it to absorb the heat. During this process, a part of the water evaporates and reduces temperature through evaporative cooling.

Cooling load (coil load)

Cooling load is a measure of the amount of cooling that is transferred during a heat exchange process. The term *cooling load* ***demand*** is used to denote the amount of cooling that is required to be supplied to a room or zone that is air conditioned. The term *cooling load* ***supply*** denotes the actual cooling that is supplied by the system. If the cooling is supplied by a coil carrying chilled water, the cooling load is the amount of heat absorbed by the chilled water per unit time. Since cooling load is the unit of energy divided by time, it can be expressed in Watts or kilowatts. But it should be noted that the cooling load does not denote the amount of electrical power consumed by the chiller; it is the amount of heat that is exchanged per unit time. Refrigeration ton (RT) is another commonly used unit for cooling load. 1 RT is equal to 3.517 kW.

The amount of heat absorbed is calculated using the well-known formula: mass multiplied by specific heat multiplied by the change in temperature ΔT. Since the cooling load has a unit of power and not energy (it is energy per unit time), the mass should be replaced by the mass flow rate, that is, the mass flowing per unit time. Mass flow rate is equal to volumetric flow rate multiplied by density. Therefore, cooling load supplied by a coil circulating chilled water is calculated using the formula:

$$Q = C_p \rho G \Delta T \qquad \textit{Eq. 4.2}$$

Where, Q is the cooling load in Watts, C_p is the specific heat of water = 4,200 J/kg/deg, ρ is the density of water = 1000 kg/m^3, G is the chilled water flow rate in m^3/s, and ΔT is the difference in temperatures of supply and return water.

Exercise 4.6 Numerical example

The chilled water flow rate in a cooling coil is 0.016 m^3/s. The temperatures of supply water and return water in the coil are 6.9°C and 9.3°C respectively. What is the cooling load of the coil?

Answer

161.28 kW.

Exercise 4.7

Using the data provided in Exercise 4.6, discuss the following:

a What happens if the flow rate is increased? Will the cooling load increase?
b What happens if the supply water temperature is increased? How will it affect the cooling load?

Answer

Even though, according to Eq 4.2, the cooling load is proportional to the flow rate, the cooling load will not necessarily increase when the flow rate is increased. The actual scenario is more complex. If the flow rate is changed, the return water temperature might change. The actual cooling supplied will depend on the characteristics of the heat exchange surface, governed by Fourier's law of conduction. If the surface is not capable of exchanging heat fast enough, the temperature difference decreases. Then, the cooling load will not increase.

Similarly, if the supply water temperature is increased, the temperature difference between the water and the material surrounding the coil will be lower. Hence the heat exchanged by the coil will reduce and the cooling load supplied will reduce.

Matching cooling load supply to demand

Every chiller has some mechanism for adjusting the load to the demand, within an operating range. In some types of chillers, the amount of cooling produced can be adapted to the demand by changing the speed of the motors driving the compressor. By running the motor faster, more cooling is produced. But there are other types of chillers in which the motors run at a constant speed. For these, the speed cannot be changed to cater to the demand. If multiple compressors are used, one or more compressors might be turned off according to the demand. For example, there are scroll chillers with 4 or 8 compressors. By turning off seven out of eight compressors, the cooling load supplied could be reduced to 12.5%. The number of compressors to be put to operation is changed dynamically depending on the demand using a control algorithm.

Coefficient of performance

The efficiency of chillers is computed using a quantity known as the coefficient of performance (COP) given by:

COP = cooling load in Watts divided by power consumed in Watts

The higher the COP, the greater is the efficiency of the chiller. Sometimes, the efficiency is also expressed in terms of kW/RT, which is the amount of power consumed for producing unit cooling. The lower this number, the higher the efficiency.

The COP depends on the operating load of the chiller, known as the part load, and usually expressed as a percentage of the maximum load. Like all mechanical equipment, the best efficiency is obtained when the load is optimal, causing the entire system to operate under ideal conditions. In most chiller types, the highest COP is at the maximum rated load.

However, there are others in which the COP decreases after a certain level. In general, the COP is low when the load in near the minimum rated value. The chiller cannot operate below a minimum load because the motors are not able to run below a minimum speed.

Exercise 4.8

The efficiency curve of a chiller is given in Table 4.1. Calculate the COP at different part loads. At what part load is the highest efficiency is obtained?

Table 4.1 Efficiency Curve of a Chiller

Chiller load (RT)	115	69	57.5	46	34.25	32.2	29.9
Chiller power (KW)	67.4	39.6	31.9	25	19.4	18.7	18.5

Answer

The efficiency can be expressed in kW/RT or COP. To calculate the COP, the RT must be converted into kW by multiplying by 3.517. Both values are tabulated in the following. The highest efficiency is at 40% part load, that is, at 46 RT.

Part Load	*100%*	*60%*	*50%*	*40%*	*29. 8%*	*28%*	*26%*
Chiller load (RT)	115	69	57.5	46	34.25	32.2	29.9
Chiller power (kW)	67.4	39.6	31.9	25	19.4	18.7	18.5
Efficiency (kW/RT)	00.59	00.57	00.55	00.54	00.57	00.58	00.62
COP	6.00	6.13	6.34	6.47	6.21	6.06	5.68

Power consumption of pumps and fans

Three types of pumps are used in a chiller system:

- Chiller pumps cause evaporation and condensation of the refrigerant
- Chilled water pumps circulate water through the cooling coils of the chiller to the AHU
- Condenser water pumps circulate water from the condenser to the cooling tower

In addition to the three types of pumps, the cooling tower has a fan for blowing air over hot water from the condenser to cause evaporative cooling. The more the amount of heat rejected into the atmosphere, the more efficient is the operation of the chiller. This is because the return water from the cooling tower is at a lower temperature and is able to absorb more heat from the condenser. When the condenser is at a lower temperature, more refrigerant condenses at lower pressure, causing an increase in efficiency.

Chilled water pumps and condenser water pumps consume less power than the chiller pump, yet they still constitute a significant percentage of the power consumption of the chiller system. Therefore, efficient operation of the pumps and fans is critical.

Since chillers operate at higher efficiency when condenser water temperature is lowered, oversizing the cooling towers or using multiple cooling towers for one chiller might be beneficial. However, the net power consumption should be evaluated to compute overall energy savings. If cooling towers consume more power than the savings obtained by increasing the efficiency of the chillers, there may not be any net savings.

Exercise 4.9

The part-load efficiency curves of a chiller for two different condenser water temperatures (CWT) are shown in the table that follows:

Chiller load (RT)	115	69	57.5	46	34.25	32.2	29.9
Efficiency (kW/RT) at CWT 30°C	0.59	0.57	0.55	0.54	0.57	0.58	0.62
Efficiency (kW/RT) at CWT 29°C	0.55	0.53	0.52	0.51	0.54	0.55	0.60

With one cooling tower consuming 2.2 kW power, the condenser water temperature is reduced to 30°C under certain conditions. Two cooling towers are needed to lower the condenser water temperature to 29°C. Discuss whether there is any energy savings in using two cooling towers.

Solution

Calculating the total power consumption of all the components, it is seen that two cooling towers are beneficial when the load of the chiller is 46 RT or higher.

Air-handling unit (AHU)

Air-handling units treat outdoor air as well as recirculated air such that its temperature and other parameters are satisfactory. A fresh-air fan supplies outdoor air to the AHU. Optionally, a return-air fan pumps air from the air-conditioned zone back to the AHU for recirculation. The fresh air and recirculated air are mixed and blown over the cooling coil carrying chilled water from the chiller. The chilled water reduces the temperature of the air and in the process, removes humidity from the air.

See the air-handling unit is shown in Figure 4.7. There are two pipes, one for chilled-water supply, the other for chilled-water return. The chilled water that is supplied from the chiller is

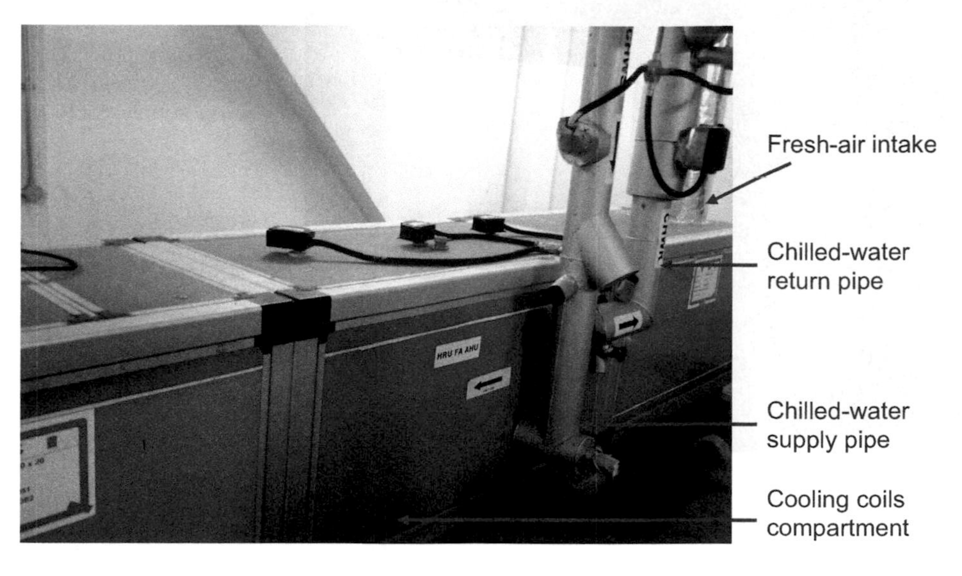

Figure 4.7 Air-handling unit

passed through coils inside the AHU compartment. There are two air ducts: intake and supply. The intake duct brings in fresh air from outside as well as recirculated air from the air-conditioned zones into the AHU compartment. There are filters for removing particulate matter such as dust and pollutants inside the compartment. The filtered air is blown over the coils. The treated air is pumped into the air-conditioned zone using a fan (See Figure 4.8).

The air that is supplied to the air-conditioned zones should satisfy requirements of temperature, humidity, and air quality. The air supply temperature should be such that after mixing the supplied air with the air in the room, the resultant temperature should match the set point. Due to heat gain in the supply duct, the temperature of air supplied to the room will be higher than that at the AHU. There is also heat gain due to internal and external loads in the room. Energy efficiency can be improved by minimizing these losses. The cooling provided by the supply air after all the losses should match the heat gain in the room in order to maintain a steady temperature.

The sensation of thermal comfort depends to a large extent on the relative humidity (RH) of air. RH value of below 60% is usually recommended. In most air conditioning systems, the primary means of achieving required RH is by cooling the air below the dew point, at which, moisture in the air starts condensing to liquid state. Usually, air needs to be cooled below 12°C to achieve reasonable dehumidification. Even though the temperature set point in the room is around 24°C, the air needs to be cooled to much lower temperatures for achieving the target RH. This has two issues. First, it requires more energy. Second, when air at such low temperature is supplied to rooms, even at the minimum flow rate specified in the codes to meet air quality requirements, the cooling provided will be more than the heat load; thus the temperature of the rooms will be lower than the set point. Many air-conditioned rooms are uncomfortable because of this. The air flow rate cannot be reduced below a minimum value because of fresh air requirements; the concentration of pollutants in the rooms is usually reduced by increasing fresh air supply. Building codes specify the minimum fresh

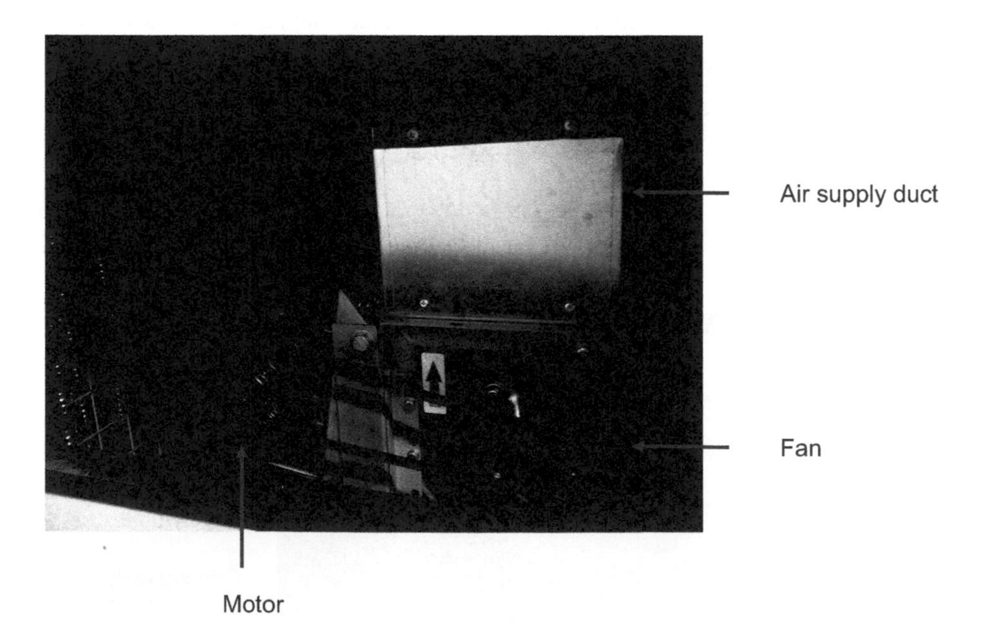

Figure 4.8 Air supply duct inside an AHU compartment

Figure 4.9 Fresh air damper

air to be supplied based on the number of occupants or the floor area. Human occupants need about 6 litres per second of fresh air for health and hygiene. Achieving the set points on temperature and RH is sometimes a challenge. See Section 4.3.3 for a possible solution.

Air that is exhaled by occupants contain high amount of carbon dioxide (CO_2). If this air is recirculated many times, the level of CO_2 in the air keeps rising. Codes specify 800 to 1,000 ppm (parts per million) as the limit for CO_2 in the air. The CO_2 level can be reduced by increasing the percentage of fresh air in the air supplied by the AHU. The concept called *demand-controlled ventilation* is gaining prominence. The main principle is to determine the amount of fresh air required dynamically based on CO_2 measurements. CO_2 is an indicator of occupancy in the room and is a marker for other pollutants and bio-effluents. If the CO_2 levels are high, the fresh air damper in the AHU opens, bringing in more fresh air from outside. A damper installed at the inlet of a fresh air duct is shown in Figure 4.9.

There are two types of systems used in air supply. The older system supplies air at constant volume, it is called Constant Air Volume (CAV). In this system, the cooling load can be met only by changing the temperature of the air supply or switching off the air supply completely. More flexibility is possible through the second system called Variable Air Volume (VAV). In the VAV system, there are dampers installed on the diffusers that supply air to the rooms. The dampers open or close, regulating the amount of air supplied to the rooms. VAV systems work best with variable speed fans. Fans with variable speed allow the modulation of air supply according to demand. In addition to meeting the demand, the power consumption of the fans is also reduced at lower speed. The speed of the fans is controlled by measuring the air pressure, which changes when dampers are opened and closed. Even though pumps with variable speed are common for chiller systems, variable speed drives for AHU fans are still not common because of the cost and complexity.

4.3.2 Control systems in air conditioning

The goal of the control system is to ensure that the required amount of cold air is supplied to the rooms such that the temperature is close to the set point while satisfying other requirements related to air quality and thermal comfort. The control system meets this objective through a cascade of events that get propagated from one subsystem to the next. The sequence of events probably starts with a sensor in the room recording a temperature higher than the set point. The VAV system controller determines the required percentage of damper opening. A command is sent to the damper actuator. When the damper opens, more cold air comes to the room from the AHU. More air flow causes a drop in the pressure in the duct. The pressure sensor in the AHU detects the pressure drop. The fan controller sends a command to the fan to increase the speed. As the air flow increases, it absorbs more cooling from the coil, causing an increase in the cooling load of the coil. This causes an increase in the temperature of the chilled water circulating in the coils. When the temperature of return water from the AHU coil to the chiller increases, the controller realizes that more chilled water needs to be pumped. It increases the chilled-water flow rate. High chilled-water flow rate increases the cooling load of the chiller. The speed of the compressor motor is increased to cater to the increased demand. All these events take place naturally without the need for a control system that integrates all the subsystems. Each subsystem works on the basis of one sensor controlling one actuator. As examples, one temperature sensor controls one VAV damper, one pressure sensor controls the speed of one span, etc. This makes it easy to implement the control systems.

While the this control system works well, better performance is obtained using high-level control of all the subsystems. This section discusses some of the ways in which energy could be saved using intelligent control.

Raising chilled-water supply temperature

The chilled-water supply temperature is usually decided at the design stage, considering all the performance parameters. This temperature of water leaving the chiller should be low enough (around 5°C) so that the temperature of the water supplied to the coils in the AHU is sufficient to cool the air below the dew point for effective dehumidification. This is a general design principle. However, there might be situations when we do not need dehumidification. There might be dry days when the RH is already at comfortable levels. In such situations, it is possible to increase the temperature of the chilled-water supply. Thermodynamically, higher efficiency is achieved by increasing the chilled-water supply temperature. For instance, it takes more energy to reduce the temperature by 1°C at 6°C than at 20°C. At lower temperatures, it becomes increasingly hard to remove the heat. In addition, to increased efficiency, the overall energy consumption is reduced by increasing the supply temperature because of the lower value of the temperature difference ΔT.

Chilled-water supply temperature could be adapted to environmental conditions. However, this is rarely done in practice because it requires integrated control. The chiller should take additional input from sensors in the AHU and occupant zone to determine the RH in real time and use the feedback to take control actions. Smart control is needed to implement this.

Optimization of the chiller sequence

When there are multiple chillers, the chillers that need to be operated at any time should be carefully selected to minimize the energy consumption. Optimization of energy through chiller sequencing is possible because the efficiency of chillers varies with the load. If chillers have different capacities and efficiencies, the optimal combination of chillers could be determined in real time by accurately estimating the load and power consumption. However, this is unlikely to be effective using design data and test data supplied by chiller manufacturers. Actual performance of chillers depends on number of factors, including the level of maintenance and site conditions. Data related to actual performance of chillers should be collected through the BMS, and this should be used for decisions related to chiller sequencing. For good performance, data should be accurate and the algorithms for data analytics should be smart. Good data analytics might also provide information about the reasons for low performance under specific conditions. Requirements for maintenance, repair, and replacement could be predicted in advance through data analytics using sophisticated machine-learning models.

Exercise 4.10

Multiple chillers of 115 RT capacity are available in a facility. The power consumption of the chiller and the pumps under different part loads was compiled from past data and is given in the following table:

Chiller load (RT)	115	69	57.5	46	34.25	32.2	29.9
Chiller power (kW)	67.4	39.6	31.9	25	19.4	18.7	18.5
Power of chilled water pump (kW)	3.0	1.6	1.2	1.2	1.2	1.2	1.2

One chiller is operated with one chilled-water pump, one condenser water pump, and one cooling tower. The condenser water pump runs at constant speed and has a power consumption of 3.0 kW. The cooling tower has a constant power consumption of 2.2 kW. What is the optimum number of chillers to be operated when the load is 92 RT? The power for the chiller and for the pumps can be interpolated from the given data.

HINT

Several combinations are possible. For example,

- One chiller operating at 92 RT part load
- Two chillers, each operating at 46 RT part load
- Three chillers operating at different part loads

For each combination, calculate the total power consumption and select the best.

4.3.3 Heat recovery systems

Several heat recovery systems are already commercially available. A common method is to extract cooling from return air and use it to pre-cool outdoor air. In many cases, return air from rooms is not recirculated because of the contaminants in the air. For example, return air from operating theatres might contain pathogens, and recirculating this air has the risk of disease transmission. The return air temperature is usually very low in the operating theatres, and there is good potential to recover this cooling. Recirculation is not usually implemented in large auditoriums and theatres because the fresh air requirement is high. In such cases, return air at low temperatures is thrown away, and heat recovery has much potential. In general, a heat exchanger absorbs cooling from the return air and transfers it to warmer air from outside.

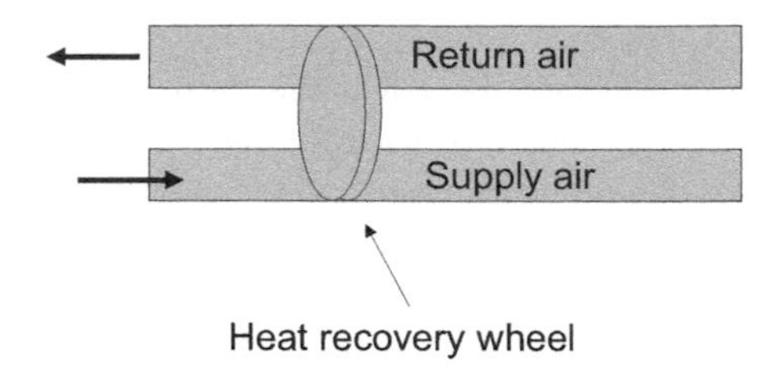

Figure 4.10 Energy recovery wheel

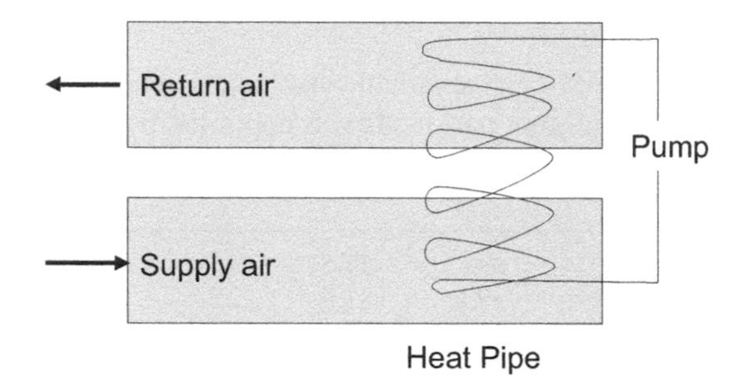

Figure 4.11 Heat pump

One method to absorb heat from return air is using an energy recovery wheel. See Figure 4.10. A wheel made of material with high thermal conductivity rotates and comes into contact with cold air. The cold air reduces the temperature of the wheel. When the wheel rotates, the colder part gets into contact with the warm air taken from outside. This cools the intake air before it is fed into the AHU. Since the temperature of the intake air is now lower, less energy is required to cool it further. It is to be noted that the wheel need not be physically in contact with air. The heat could be exchanged using an appropriate medium so that contaminants from the return air are not transported to the fresh air.

Another concept is a heat pipe or heat pump. See Figure 4.11. A liquid or refrigerant is circulated in a pipe, which absorbs the cooling from the return air duct and transfers it to the air in the supply duct. In other words, heat is pumped from the intake air to the return air.

These solutions can be used to reduce the energy required to cool fresh air significantly because of the reduction in the temperature of intake air. These solutions could also be used to improve thermal comfort by preventing overcooling of rooms where a large volume of fresh air is required, and the temperature of dehumidified air is low. Heat exchange between dehumidified air and the intake air raises the supply air temperature to comfortable levels. No active energy is used to heat the supply air, and the intake air is also precooled.

Outdoor air can be pre-cooled by geothermal systems, evaporative cooling using spray ponds, and other techniques. Geothermal systems work well in cold climates where the ground temperature in summer is much lower than ambient air temperature. Because of the thermal mass of earth, the temperature deep underground remains low even when the air temperature increases in summer. This cooling can be absorbed and pumped using an appropriate mechanism and transferred to the intake air. Since these systems require additional energy to operate the heat pumps, they are effective only under certain conditions. The control system should be smart enough to decide whether to use pre-cooling, depending on outdoor and environmental factors.

4.3.4 Reducing the cooling load

The most effective way to reduce the energy required for air conditioning is by reducing the cooling load demand. An "Energy Smart" building should use minimum amount of energy without compromising on user comfort. While reducing the cooling load demand, care should be taken that thermal comfort parameters are not compromised. In fact, energy and comfort are conflicting objectives that have to be carefully optimized in design and operations (Pantelic et al., 2012).

The cooling load demand depends on the heat ingress into the air-conditioned space due to internal and external sources. The external source is primarily the sun and the objects heated by the sun. Internal heat sources include people and equipment. Heat load consists of two parts: sensible and latent. ***Sensible heat*** causes change in temperature. ***Latent load*** is due to the moisture in the air, which causes heat to be released during condensation. Various components of the heat load are discussed in the following sections.

Solar heat gain

External heat ingress is primarily from the radiant heat through window glass, conduction through the envelope, and infiltration through openings in the building. Heat transmitted into the building through conduction, convection, and radiation should be reduced to minimize energy required for cooling the building.

Conduction is governed by the Fourier's law:

$$Q = \frac{k\,A\,\Delta T}{s} \qquad \textit{Eq 4.3}$$

Where,

Q = heat transferred per unit time (W)
A = area of cross section through which heat is conducted (m^2)
k = thermal conductivity of the material (W/m K)
ΔT = temperature difference (K)
s = material thickness (m)

In the case of walls and windows, heat is conducted through the thickness of the element from outside to inside; therefore, A is the area of the surface. Thermal conductance (also known as thermal transmittance or U value) is the heat transmitted through unit area of the surface, that is,

$$\text{Conductance } U = k/s \ (W/m^2K) \qquad \textit{Eq 4.4}$$

Knowing the U value, the heat transmitted per unit time is calculated by multiplying it with the surface area and temperature difference. The thermal resistance is defined as,

$$\text{Resistance } R = 1/U \qquad \textit{Eq 4.5}$$

If the material is made of multiple layers, the net resistance is calculated as the sum of the resistances of each layer in the path. These resistances are considered as connected in series. If there are multiple parallel paths for heat transfer through, it is equivalent to resistances connected in parallel. In that case, the net heat conducted is equal to the sum of the heat transmitted through each path.

For reducing the heat transmitted by conduction, the following strategies will help:

a Reduce the area of thin elements such as window glass
b Increase the thickness of walls
c Use materials with low thermal conductivity (high thermal insulation)
d Reduce the temperature difference between the outer and inner surfaces – for example, shade external walls so that the temperature of the outer surface can be reduced

Radiative heat transfer is governed by the Stefan–Boltzmann law:

$$Q = \varepsilon \sigma A T^4 \qquad \textit{Eq 4.6}$$

where,

ε = emissivity of the body
σ = Stefan–Boltzmann constant
A = area
T = temperature

Radiant heat comes into buildings through glass windows. The most significant component of radiant heat is, of course, that from the sun. Radiation from the ground is also significant in hot climates. Hard ground absorbs solar radiation and gets hot, radiating this heat into the interior of buildings. By replacing hard ground with material of low emissivity, such as grass, this component of heat can be reduced. Heat radiated by objects in the vicinity of the building contributes a small amount because the temperatures of these objects are not very high.

The solar thermal transmission through window glass is dependent on the properties of the glass, in particular, a property called Solar Heat Gain Coefficient (SHGC). It is the fraction of solar radiation transmitted through the glass. Shading Coefficient (SC) is another parameter commonly used to characterize solar transmission through glass. It is the ratio of the SHGC of the particular type of glass and that of pure glass. SC is usually correlated to the visible light transmittance of the glass. Glass having high visible-light transmittance also tends to transmit a high amount of solar heat. This is because radiant heat and light are part of the solar spectrum; they differ only in the wavelength. However, advances have been made in selective transmittance of light using special coatings that are sensitive to the wavelength of light. There are coatings that absorb or reflect infrared but transmit visible light. Such spectrally sensitive materials help to reduce the radiant heat, while improving daylighting.

Convective heat transfer into buildings happens by the movement of air either through infiltration or by hot air absorbing cooling when it comes into contact with cold surfaces like windows. Gaps in windows and doors allow infiltration of hot air into the rooms. In addition to contributing to sensible heat load, it also adds moisture to the air, which adds to the latent load. Automated doors and windows might help to ensure that they are kept open for the minimum time required.

Thermal simulation software such as EnergyPlus (Crawley et al., 2000) could be used to compute the heat ingress into buildings. However, this requires a detailed modeling of the geometry and material properties; this is usually time-consuming. Approximate methods have been developed. For example, a concept called ETTV (Envelope Thermal Transmission Value) has been developed. Rough estimates of solar heat gain can be obtained from a simple formula without detailed geometric models. ETTV is expressed in W/m^2 and represents the average heat transmitted per unit area of the building envelope. Multiplying by the area of the façade gives the average heat transmitted through the façade.

Internal heat gain

All the power consumed by electrical equipment inside a building usually gets converted into heat. This adds to the heat load of the building, unless this heat is somehow removed from the equipment before it is absorbed by the air in the occupied zones. Engineers have attempted clever ways of preventing heat from equipment adding to the cooling load. For example, heat dissipating parts of LED lights are put above the false ceiling and this heat is removed by fans. Large data centers use chilled water to cool the processors of computers so that this heat does not get absorbed by the air inside the building. Where such techniques are not used, the heat load from equipment should be calculated by summing up the rated power of all the electrical devices used in the building.

People contribute significantly to the internal heat load. The heat dissipated by people depends on the metabolic activities as well as physical characteristics. An average person in a sedentary position contributes about 65 Watts of sensible heat and 65 Watts of latent heat. Latent heat load is due to the moisture exhaled by people during breathing. While cooling the air, the moisture condenses and releases heat, thus contributing to the cooling load. By counting the number of occupants in the room, the heat load due to people can be estimated. In addition to the latent load from people, the cooling load constitutes a component called the *ventilation load.* It is related to the fresh air required by people and is proportional to the number of occupants. The fresh air taken from outside should be cooled and moisture must be removed through condensation. The heat removed in the process can be calculated using psychrometric charts. Psychrometric charts give the relationship between temperature and humidity in air. By estimating the amount of humidity that must be removed from air while cooling the air, from the initial temperature to the desired temperature, the cooling load can be calculated.

Reducing heat gain through automation

Many actively controlled devices have been designed to reduce the heat load in buildings. A simple device that can be actively controlled is a window shade. By shading the window, direct solar radiation into the rooms can be prevented. It will also help to reduce the heat transmitted by conduction by preventing the window glass from heating up, thus reducing the temperature difference. Walls can similarly be shaded to reduce heat conduction. However, shades prevent daylight transmission into buildings. Shading devices such as window blinds prevent visibility as well; people do not prefer these. Shades can be made more effective by adaptive control. Rotatable window shades, roller blinds, and similar devices have been successfully implemented in many buildings.

Electrochromic glass is another solution for reducing radiant heat transfer. The transmittance of the glass is varied by applying a voltage to a filler material that changes its properties. When it is too hot, the glass is made opaque, and it blocks heat. At other times, it is made transparent. Thermochromic glasses are also commercially available. The transmittance of the glass is varied according to the temperature.

Automated lighting control helps in reducing the heat load. Since electrical lights generate heat, if lights are dimmed or turned off when they are not needed, it will reduce the cooling load as well.

The ventilation load can be reduced by accurately determining the number of occupants. This strategy is called as demand-controlled ventilation. The fresh-air intake is adjusted according to occupancy, and additional energy required to remove the sensible and latent load in the fresh air is reduced.

4.3.5 Dehumidification

In hot and humid climates, dehumidification constitutes about 50% of the cooling load. This is the cooling required to absorb the latent heat given out during the condensation of moisture in the air. There are dehumidification systems that reduces the RH of intake air without the use of active energy. If air is dehumidified before passing to the cooling coil, the energy consumption can be reduced. Solid and liquid desiccants have been used to absorb moisture from the air. Concentrated salt solutions have been used as liquid desiccants (Kasamsetty and Raphael, 2018), these have high affinity for water. When moist air is passed through concentrated salt solutions, the moisture is absorbed, and dry air comes out. However, there are two issues. First the moisture that is absorbed causes the liquid to be diluted and it becomes less effective in absorbing moisture after some time. The liquid needs to be regenerated by removing the moisture. Using solar energy for this is an option. Second, the latent heat that is released during the absorption of the moisture causes the temperature of the solution to rise. The system should be designed to remove the heat generated so that the air temperature does not rise. A carefully designed, actively controlled dehumidification system has the potential to decrease energy consumption significantly.

4.3.6 Thermal energy storage (TES)

The concept of TES involves storing cooling energy in some form such as chilled water, ice, or phase change materials. The stored cooling could be used later when required. Currently many TES applications are designed to benefit from lower energy tariff during off-peak demand periods. The idea is to use energy to produce cooling when the energy tariff is lower. The cooling is stored in the form of chilled water or ice. The stored cooling is used to cool the intake air when the demand is high.

TES can also be used to operate the chiller at the optimal load. The chiller produces excess chilled water when the load is below the optimal load. The excess chilled water is stored and used when the cooling load is high. The storage and release of chilled water is decided through automated control. Smart control should take into account storage losses and predicted future demand.

4.3.7 Personalized ventilation

Conventional air conditioning systems aim to cool the entire volume of the occupied space. Air is supplied to ensure that the entire zone is maintained at set point temperature and the treated air satisfies air quality requirements everywhere. An alternative is to cool and condition the air only in the immediate neighborhood of occupants. This is effective in conditions such as offices, where occupants remain at their seats most of the time and do not move a lot. In that case, energy can be saved by restricting cooling to a smaller volume of the room. This is achieved by using personalized ventilation (Chen et al., 2012). A diffuser is placed on the desk that brings fresh air directly to the breathing zone of the occupant. Since this air is not mixed with ambient air, the air quality is also improved. Drawbacks of this include the requirement for additional ductwork to bring fresh air to the desk, pressure drop due to longer ducts, obstructions on the desk due to the presence of the diffuser, etc. Many of these issues can be eliminated using careful design.

4.3.8 Radiant cooling

Radiant cooling is an alternative to conventional air conditioning in which the cooling is provided through the supply of cold air. In radiant cooling, cold panels are used to reduce the temperature primarily by radiant heat transfer. Typically, surfaces with large area like slabs and walls are cooled by circulating chilled water in pipes which are in contact these surfaces. These in turn radiate cooling (absorbing heat through radiant heat transfer). In addition to radiant heat transfer, it is inevitable that a certain amount of cooling takes place by convection because air absorbs the cooling and gets mixed up with the ambient air in the room. Since the radiant system does not supply fresh air, dedicated outdoor air systems are usually provided in addition to radiant panels.

Where fresh air requirements are low, radiant panels are more energy efficient because they avoid the additional step of cooling and transporting air. Cooling is directly supplied to the rooms through chilled-water pipes. Pumping chilled water is more efficient than pumping cold air. However, radiant systems should be carefully designed to avoid problems such as condensation on the pipes and panels, leaks, high relative humidity in the rooms, etc.

Radiant systems are best suited for external walls in which the solar heat absorbed by the wall is immediately removed through radiant pipes. By cooling these walls, the heat is prevented from getting transferred to the conditioned air; cooling this air is more expensive in terms of energy consumption. Window sills and internal light shelves can also be cooled by radiant pipes such that radiant heat absorbed by them is immediately removed before it enters the air-conditioned zone.

4.4 Energy management

The previous sections discussed ways to reduce the energy consumption for lighting and thermal comfort. In addition to managing the demand, building owners are equally interested in optimal production of energy. Today, many buildings have alternate sources of energy, including solar and biomass. Since the alternate energy sources require considerable investment, it is essential to maximize the production and ensure optimal use of this energy.

Solar energy is commonly tapped using photovoltaic (PV) cells. Many tropical countries receive abundant solar energy. Peak power of about 1,300 Watt per square meter of solar energy reaches the surface of the earth. Out of this, only 42% is in the visible spectrum. About 50% is heat (infrared). Most PV cells absorb power only from the visible spectrum and have efficiency of less than 20%. Therefore, only about 100 Watts per square meter of electricity might be generated from solar cells. Since solar panels are typically installed on rooftops, and usually this space is limited, optimal use of the solar panel area is required. The cost of solar panels and the cost of usable space should also be accounted for in the calculation of benefits.

How do we maximize solar power production with the available area of solar cells? Several factors should be considered. Power production is maximum when the sunrays reach the surface of solar panels at right angles. This is because of two effects. First, only the area of the panel that is perpendicular to the path of the rays will be effective in blocking the radiation and absorbing this energy. Therefore, if the sun rays have a non-zero incident angle, the power produced will be reduced proportionally to the cosine of this angle. Next, the effectiveness of knocking out electrons from the photovoltaic material depends on the incident angle; the best efficiency is obtained when the rays are perpendicular to the surface. For certain types of solar cells, angles more than 45° are not effective in capturing solar energy. More solar radiation is lost by reflection from the surface of the solar panels when the incidence angle is increased.

To maximize power production, the solar panels should be ideally normal to the path of sunrays. However, the position of the sun is not fixed to ensure that the panels are always at 90° with respect to the incoming rays. Apart from the movement of the sun from east to west during the day, the position of the sun also shifts from north to south and back again during the year. The position of the sun for a specific latitude and longitude is available in the solar path diagram for the location. The position of the sun at any time is obtained from this diagram. By studying this diagram, the optimal orientation of the panels might be determined by calculating the average performance over the whole year.

We can rely on automation to improve the efficiency of solar panels further. Using solar trackers, the orientation of the panels could be changed according to the position of the sun. Single-axis and dual-axis solar trackers are commercially available. Photosensors in combination with stepper motors could be used to control the horizontal and vertical angles of the panels.

Solar power production can be increased by capturing energy from a larger area without increasing the area of solar panels. This is done by using reflectors. Mirrors could be used to redirect light to fall on the solar panels. Since the cost of reflectors is less than the cost of solar panels, it is a less expensive way of increasing solar power production. The reflectors could be rotated in response to the position of the sun for maximizing the amount of light falling on the panels.

One critical issue is that solar energy is available only during daytime, and the electricity production varies with sky conditions. Steady power is not available throughout the day, and no power is produced at night. Two strategies are used to address this issue. The first is integrating the photovoltaic system with the electricity grid. Any deficiency in PV production is automatically managed with grid energy. The other possibility is to store PV energy in batteries and meet all the demand from batteries. However, batteries are not environmentally friendly due to the chemicals used in their production. When more environmentally friendly technologies become available for energy storage, sustainability of solar energy production will be improved. The discussion on thermal energy storage in Section 4.3.4 is relevant in this context.

In addition to solar PV, solar thermal systems are now being increasingly used. Solar thermal systems absorb heat from the sun to produce usable energy. Large reflectors are used to concentrate solar power to a small area to increase the intensity of power and to increase the efficiency of energy production. The concentrated energy could be used to boil water to run turbines and produce electricity. Without the use of solar concentrators, it is difficult to produce high temperatures. Resulting "low grade" heat cannot be used effectively. However, where there is large demand for hot water, in residences and hospitals for example, simple solar water heaters are useful. Solar water heaters have panels to trap solar heat, which is absorbed by circulating water in small pipes. The panels should be oriented in the direction perpendicular to the sun for the best efficiency. Automation techniques that were discussed for solar PV panels are equally applicable for solar water heaters as well. If hot water demand is high only in winter, the orientation can be optimized by considering the solar path during the cold months alone; this normally gives a much narrower range of angles compared to considering the sun path for the entire year. In that case, it may be enough to rotate the panels about a single axis, and the cost of installation can be reduced. Static solar water heaters work efficiently only during limited hours of the day when the panels are perpendicular to the radiation from the sun.

4.5 Monitoring building performance

The most effective way to enhance building performance is through management control. This requires active monitoring and measurements, as well as timely corrective actions. First, there should be targets on performance parameters such as energy. This requires developing an energy model with realistic data. Then, we need to closely monitor the energy consumption and take corrective actions if deviations from expected values are noticed. This cannot be done without sufficient level of details in the modeling and measurement. The energy model should predict the energy consumption of each building system, not just the overall consumption of the building. There should be submetering to monitor the consumption of each building system individually in real time. Environmental parameters should also be measured so that smart analytics could be carried out to find out the factors that cause large deviations in model predictions.

Sensing, data analytics, and automated control are integral part of a strategy for a high-performing building. This author was part of the team that implemented this strategy in the first net zero building in Singapore at the BCA Academy. More than 600 sensors were deployed in this building, and a custom dashboard was developed to monitor the performance in real time. Real-time measurements helped to identify systems that were functioning well according to design and to take rectifying actions where assumptions were not fully accurate.

Many commercial building management systems have intuitive graphical user interfaces that help operators quickly view and analyze the performance of all the systems. A screenshot of the BMS of an experimental facility is shown in Figure 4.12. This shows the schematic of an air-handling unit (AHU). The current cooling load of the AHU is shown at the bottom (3 RT). The set points and other status variables are also displayed. The operator can change the set point by clicking on the appropriate controls and inputting values. The air temperature at various stages within the duct are visible. The intake air from outdoors is at 32.1°C and has an RH of 72.9%. Its temperature reduces to 25.5°C due to pre-cooling through the heat pipe. Then it further reduces to 12.5°C after passing through the cooling coil. After the second leg of the heat pipe, the temperature increases to 17.4°C due to heat exchange with intake air. The final supply temperature to the room is 21.4°C. All this data can be visually analyzed by the operator, and any potential problems could be detected. For example, if the off-coil temperature (temperature of air after passing through the cooling coil) is high, the operator might suspect that the chilled water might not be at the right temperature. He could then examine the interface of the chilled water system and identify the problem. The schematic

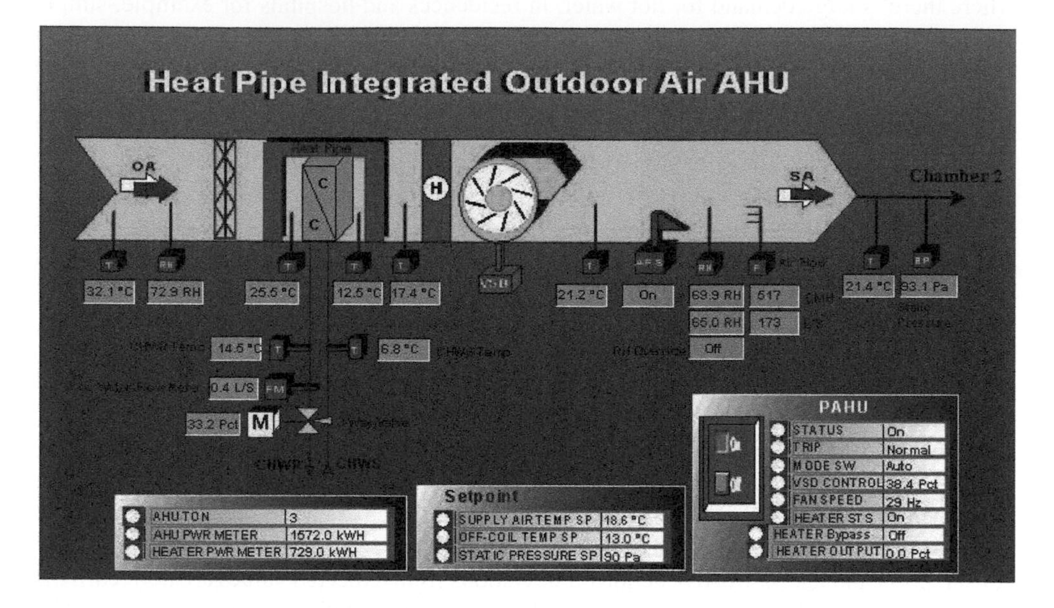

Figure 4.12 Screenshot of a BMS

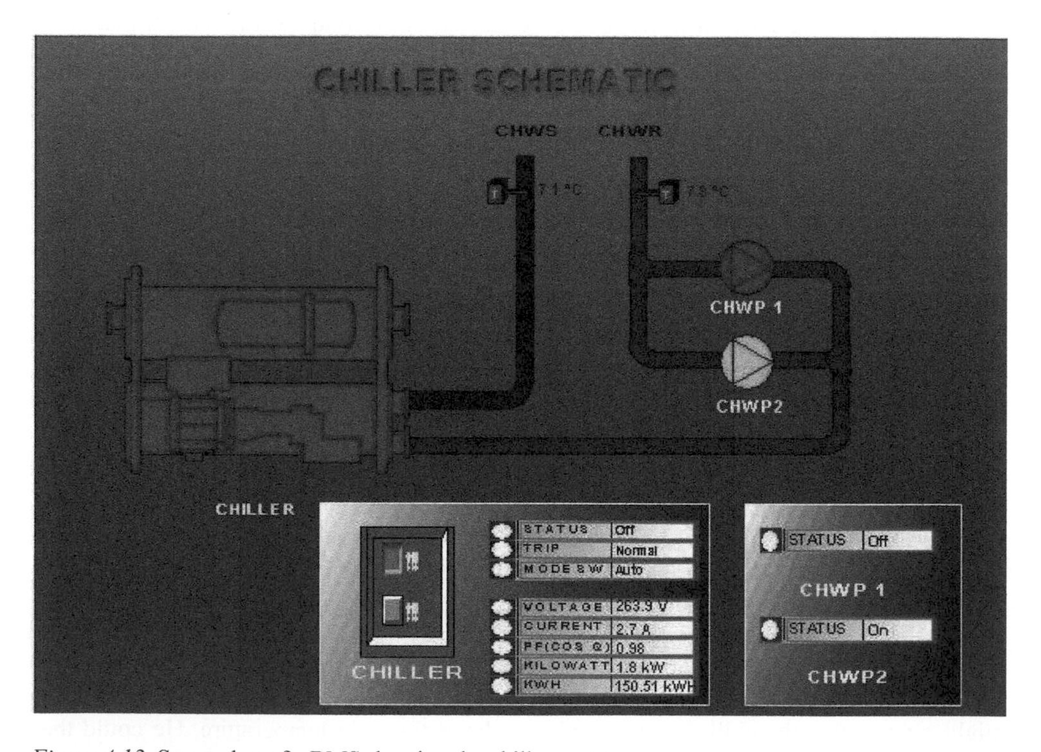

Figure 4.13 Screenshot of a BMS showing the chiller system

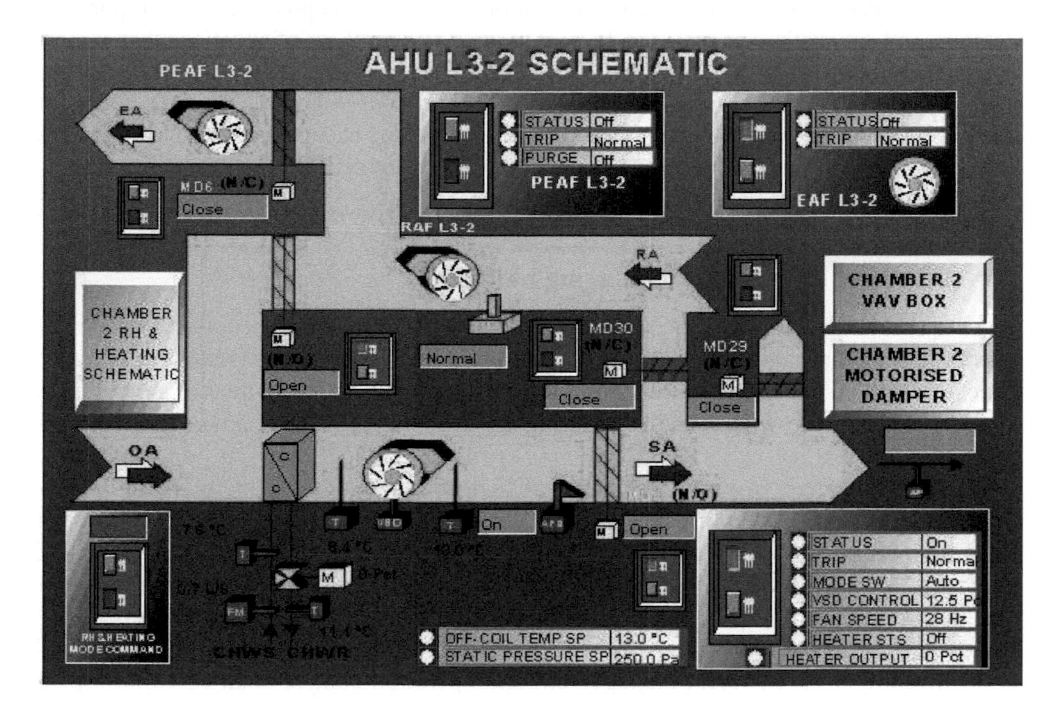

Figure 4.14 Screenshot of a BMS for an AHU that has a return air duct

of the chiller system is shown in Figure 4.13. The supply and return water temperature, the status of pumps, etc. are available in this schematic.

A more complex AHU system is shown in Figure 4.14. The complexity of this system is increased because of return air duct for recirculated air. There are a number of dampers that control the flow of air under different situations. Understanding such a complex system might be challenging for novice engineers; however, these are quite convenient representations for experienced air conditioning engineers.

For better effectiveness and convenience, there should be a dashboard for the entire building, not just for the air conditioning system. This helps to monitor the performance of all the building systems from a single window. The dashboard should permit performing different types of analytics using sensor data, and display the conclusions graphically. Any intervention should be possible directly through the interface. This requires tight integration between all the building systems.

4.6 Safety and security

The most successful deployments of building automation have been in the area of safety and security. Since "high-net-worth individuals" give high priority to security, lot of investments have been made in security systems, and security devices have been commercially successful. CCTV and IP cameras have become very sophisticated with image recognition, motion detection, and smart recording. Smart cameras automatically zoom in on areas where suspicious activities are detected. Machine-learning techniques have been used to detect anomalous scenes and unsafe conditions. Remote monitoring is possible through portable devices, and automatic messages can be sent to users to alert them of situations that require personal attention.

Technologies such as computer vision, voice recognition, fingerprint, and gesture have been used for access control and improving convenience of occupants. However, cybersecurity concerns of these systems have not been fully addressed. More advanced machine-learning systems might be needed to address these concerns.

4.7 High-level software for managing buildings

The low-level software that was discussed in the previous chapter is used mainly by developers who implement automation systems. End users need high-level software that does not expose too much of the complexities of the implementation. In the context of smart buildings, these users are owners, facilities managers, or operators (technicians). Their requirements are different.

In the first chapter, it was mentioned that smart buildings have computerized operating and management systems that provide control over all the services. Ideally, these must be integrated with organizational processes so that productivity and efficiency of managers, operators, and occupants are maximized. The goal of the management system is to create an environment that maximizes the efficiency of the occupants of the building while at the same time allowing effective management of resources with minimum lifetime costs. Efficient management and operation control is an essential aspect of smart buildings.

Historically, the management software for building systems is called the Building Management System (BMS). In practical usage, BMS refers mostly to the system that controls the operation of HVAC systems. The software that controls lighting is called a Lighting Management System (LMS). Similarly, there are separate systems for managing fire safety, security etc. Since integrated systems have better potential for smart autonomous behavior, it is better to bring all the separate systems under the umbrella of the BMS.

There is another piece of software that is related to the management of buildings. It is called a Facilities Management System (FMS). Traditionally, the roles of FMS and BMS are different. A rough guide to the functions of BMS and FMS is given in Figure 4.15.

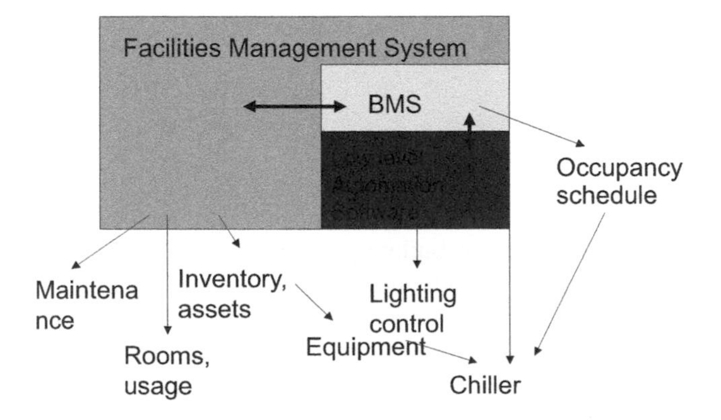

Figure 4.15 A layered view of software for building management

4.7.1 BMS

A few screenshots of a BMS were shown in the previous section to illustrate its functions. Most centralized air conditioning systems come with such BMS for the convenient operation of the systems. However, in many buildings BMS contains only the details of air conditioning systems. Ideally, all the systems should be tightly integrated into a single software. In reality, tight integration is not possible because vendors are specialized in individual building systems; they supply software that cannot easily communicate with other systems. Therefore, we end up with a BMS that can only monitor the operations of the HVAC system and a lighting management system that can only control the lights! The situation is likely to change in the future when people become more aware of the possibilities of smarter integrated systems.

4.7.2 FMS

In this section, the requirements and features of facilities management software are discussed. It will be explained how FMS contributes towards the intelligence of buildings.

Importance of scientific approach to management

Good technology should be supplemented by proper management for attaining good performance. For example, a building might have a state-of-the-art lighting control system, but without well-defined maintenance procedures, the system breaks down in a short time. After some time, bulbs do not work, some keep flickering, etc. The building not only gives a poor impression, but also results in poor efficiency of its occupants. While the systems are getting repaired, the occupants face inconvenience, discomfort, and loss of productive time.

Exercise 4.11

Prepare a list of tasks that need to be carried out for the efficient operation of a facility such as a lecture theatre. How do you make sure that all the activities are carried out promptly and the facility remains in good working condition all the time?

Supporting management activities in an organization

Management activities belong to three levels:

- *Strategic* – setting of organizational goals and objectives
- *Tactical* – methods of implementing goals and objectives by setting of tasks
- *Operational* – methods of undertaking actual tasks

Strategic planning requires data and a user-friendly interface for visualizing and analyzing data. Whether data are collected automatically using sensors or manually input by users, managers require online and easy access to it for making strategic decisions. For example, a facilities manager might take a decision to reduce the energy consumption by 30% after analyzing the pattern of energy usage.

Tactical decisions are also made with the support of data analytics. Data might reveal that idle equipment consume significant amounts of energy when occupants are not present in the building. A tactical decision might be to encourage users to turn off the equipment when they leave the building.

Finally, at the operational level, steps could be taken that involve a combination of automation and human behavioral aspects. Displaying statistics about wasted energy in a prominent place is a means of sensitizing users about the issue.

Management activities involve cycles of planning, implementation, monitoring, and adjusting procedures. The course of actions to be taken are decided during planning. Implementation involves performing the actions by communicating to the right people or setting up the right set of processes. Whether the action has been done the right way is monitored, and whether the expected outcomes are achieved are then monitored. If the expected performance is not obtained, the procedures are modified.

All these activities require information gathering and communication. A computerized management information system (MIS) helps in taking decisions and implementing them. Performance of the system can be monitored regularly, and corrective actions could be taken quickly when faults are detected. Management decisions need to be communicated efficiently to people in charge of operations. This is best done by automating the workflow. *Workflow* refers to the flow of information in an organization. In an ideal situation, information should flow seamlessly from one person or department to the next without any physical exchange of papers. The flow of information can be tracked, and bottlenecks can be identified.

FMS is both a management information system and a workflow automation system for managing large facilities. It is an integrated software system that facilitates smooth operations of all the systems in a building complex. It permits a single-point access to coordinate all the activities needed for the smooth operation of the facility. It is also a repository of information that helps take management decisions at different levels.

Exercise 4.12 Group discussion

Discuss what types of information are needed for facilities managers and how they are used for decision-making.

Features of an FMS

Current commercial FMS have some of these features:

- Inventory management, asset management, material management
- Resource management
- Maintenance management
- Security management
- Alarm and crisis management
- Integration with building systems

4.7.2.1 Asset management

Facilities managers need to have quick access to information related to all the fixed and moveable assets in an organization. This is necessary for efficient allocation and utilization of resources.

Example: An employee needs a new projector for a meeting. The facilities manager looks up the asset management database and finds that some projectors are remaining idle and are not being used. The employee borrows the projector, and the physical movement of the asset is recorded in the database so that the employee can be contacted if the projector is needed by someone else. If tracking devices are fixed to the equipment, the physical location of the asset can be tracked in real time.

Asset management procedures differ from organization to organization. These procedures make sure that assets are properly recorded and can easily be tracked. For example, at the time of the purchase of a piece of equipment, an asset number is allocated if the cost is more than a specified value. The equipment details are entered into the FM software. These procedures should be integrated into the workflow of the organization for effective enforcement.

Inventory management is part of asset management. A facility might have one or more stores where spare parts of equipment and other materials and consumables are kept. The facilities manager needs to know what materials are available in the store, how old they are, and how much quantity is available. Inventory management subsystem of an FMS caters to these requirements. Procedures should be defined to keep the inventory up to date, replenish spares, order placement, etc. There is much potential for automation of these tasks.

TRACKING TECHNOLOGIES

It is best to track assets using automated techniques. Many technologies have matured and successfully demonstrated in many projects. These include,

- RFID (Radio Frequency Identification)
- Bar codes and QR codes
- Bluetooth Low Energy (BLE)
- Wireless

Barcodes are widely adopted due to low cost. A barcode is printed on paper or other material and pasted on objects that need to be tracked. A barcode contains a series of lines encoding the data which is read using a barcode scanner. This data (identifier) is usually mapped to records in a database which can be quickly retrieved by reading the barcode. QR codes are two-dimensional images serving similar function. Cameras scan the images, and the QR code software decodes the information contained in the images. QR codes can be printed easily without the need of expensive equipment.

Barcodes and QR codes have been used for tagging, and fast, accurate data entry. Information related to an asset, such as location, attributes, etc. can be stored in a database and linked to the codes in the tags. This information is easily retrieved simply by scanning the code. Hence, they are effective in fast data entry while taking out an asset or equipment.

The main advantages of barcodes and QR codes are low cost, reliability, and availability. The disadvantages include relatively short-life and requirement of line of sight for reading.

RFID

RFID overcomes most of the limitations of barcodes. They use radio frequencies to send and receive data using RFID scanners. The tags can be rewritten multiple times and can be reused. They are of two types of RFID – active and passive. Active RFID need power to transmit data to scanners. Passive RFID gets activated when RFID scanners send out an activating frequency. The range of passive RFID is usually limited to about 1 meter. Active RFID can transmit its signals to more than 10 m.

The advantages of using RFID are durability and ability to track wirelessly without requiring line of sight. They are slightly more expensive compared to barcodes.

BLE

Bluetooth Low Energy uses the familiar Bluetooth technology that is present in mobile devices. It has low power requirements and sends data in small packets. BLE devices broadcast data such as identifiers, which are read by receivers. The signal strength of transmission is used for detecting the location of the device. BLE devices are durable, portable, and cheap. However, they have limited range without line of sight.

UWB

Ultra-wideband (UWB) technology is like active RFID but using signals of a different frequency range (greater than 500 MHz). UWB has a large range (up to 1000 m) and can be used for real-time tracking. Using multiple beacons for sending and receiving signals and calculating the time of flight, locations can be identified fairly precisely through triangulation. The disadvantage of UWB is that they are relatively expensive.

WIRELESS

Connectivity using Wi-Fi is now common. Wi-Fi uses radio signals to connect to the internet through a wireless router. This technology can also be used to track devices by locating the router to which the device is connected. Signal strength has been used to identify the location of devices using this concept. Zigbee is another global standard for wireless communication and is cheaper as well as energy efficient.

Exercise 4.13 Group discussion

Many items are efficiently stacked and stored in a warehouse. For optimal use of space, an automated system is used to lift and store items in boxes that are uniquely labeled. Explain how tracking devices help in the efficient operation of the facility. Discuss what hardware is needed for implementing automated storage and retrieval of items.

4.7.2.2 Resource management

Resource management is similar to asset management. Resources include people, space, money, and equipment. A facilities management system should provide support for efficient utilization of resources. This requires enumerating all the resources in a central database and keeping track of the utilization rate. Workflow automation helps in this. However, apart from automating the flow of information, automation of physical tasks has much potential. A time-consuming activity for implementing this is developing an as-built model of the facility. Many changes take place in the building during the construction as well as during the service life; design details may not be valid during the service life. An important aspect of the as-built model is the geometry.

Laser scanning, also known as LiDAR (Light Detection and Ranging), has demonstrated potential to provide fast, accurate, and reliable geometric representation of construction facilities. It uses laser to generate a point cloud of data containing the x,y,z coordinates of points of the scanned object. Several researchers have demonstrated the use of laser scanning (Perez-Perez, 2021; Agapaki and Brilakis, 2021) and imaging techniques (Omar et al., 2018) for creating or updating building models. Despite impressive advances, there are several challenges in creating models from point cloud.

After creating an accurate model of the as-built facility, the information can be used for analyzing the utilization rates and effectiveness of usage. For example, it is useful to know how many people are using a conference room. Apart from compiling information from room booking system, data from occupancy sensors might be used for supplementing the analysis.

Today *building information modeling* (BIM) technology has become widely adopted in construction projects. The central idea is to have a complete three-dimensional representation of the building with all the attributes that can be used in various stages of the life cycle of the building, right from the conceptual design stage. However, it has not been widely used for facilities management currently. There is considerable potential for the use of BIM in FM. Imagine that facilities managers have a model of the building on their computer, and they can click on a piece of equipment, for example, the chiller, to get all the information about it, when it was installed, the make, the warranty period, where it can be serviced, etc. They can search the building model to locate all the projectors of a particular type that are installed. They can compute the floor area of any room or zone instantly. These are examples of potential benefits of BIM for FM. However, this requires accurate models of the building with high level of details. Laser scanning and related technologies might help to create as-built models and keep them up to date during the service life of the building.

Exercise 4.14

A large campus has many buildings with a network of utilities crossing them. There is a plan to construct a new drain on campus. The excavation for the drain should be carried out without causing any damage to utilities. Discuss how the FMS helps in this task.

HINT

- All the underground utilities should be clearly represented in the building information model and linked to the FMS.

4.7.3 Maintenance management

Companies attempt a variety of strategies for ensuring that systems are operating efficiently while costs are reduced. Maintenance management tasks include scheduling routine maintenance activities, predicting and preventing failures, and undertaking corrective actions (repairs). Routine maintenance could be scheduled in the workflow automation system so that the tasks get completed on time without relying on people. Predictions and preventive maintenance require more intelligence. Data analytics and machine learning could help in predicting when the equipment is likely to break down. Different types of sensors could be used to detect early signs of failure. Corrective actions could be taken automatically. For example, repair and emergency maintenance could be scheduled and the right people could be contacted. Creation of work orders, generation of the details of work to be done, allocation of the task to workers, etc. could be done automatically. When the work is completed, managers are briefed on the details, and feedback could be collected for continuous improvement of the system. These involve tight integration between the building management system and the workflow system in the FMS.

Smart maintenance management software helps in reduction in labor costs, ensures increased equipment availability due to better planning, increased equipment reliability through identification of repetitive faults, and optimal stocks of spares and materials.

Exercise 4.15

A maintenance management system accepts requests from users of a building for fixing faults in electrical, mechanical, and civil systems. The requests are routed automatically to the maintenance overseer of the specific system. The location of the faulty system should be accurately input into the software so that the right persons are identified, and they are directed to the precise location. Determine the features that are needed in the FMS to facilitate this.

HINTS

- Building information model (BIM) integrated with GIS (geographic information system) could be used to input the coordinates of the location.
- Information about personnel and their responsibilities (roles) could be stored in the workflow automation system.
- Users and their work locations might also be available in the workflow automation system.

Exercise 4.16

A company managing a large estate of residential buildings has given several buildings on lease. When a tenant vacates, the company should decide what repair and maintenance work must be carried out before the next tenant arrives. How does the facilities management system help the company in performing this task?

HINTS

- A history of previous work performed in each building should be appropriately stored in the database of the FMS.
- This data should be linked to individual building elements and equipment.

4.7.4 Security management

Role-based security privileges could be assigned in the FMS. The list of occupants of the facilities, where they have access, the hours of usage, etc., could be entered in the FMS.

If the FMS is linked to the building management system, access control systems could be used to enforce the rights. Unauthorized entry can be prevented, and all movements can be tracked.

4.7.5 Alarm and crisis management

Different types of crises could develop in large facilities. This includes security breaches, fire, accidents, etc. Facilities management could help in managing crisis. Crucial information needed for planning operations could be instantly obtained from the FMS. Processes could be quickly set in place, and actions could be performed smoothly through automated systems facilitated by the integrated workflow and building management systems. Smart response to critical situations is a hallmark of intelligent buildings.

4.8 Summary

Automation has tremendous potential to improve the performance of building systems. A number of examples of automated building systems are presented in this chapter. Energy consumption of lighting and air conditioning systems can be reduced significantly through automation. Using light sensors and occupancy sensors, electrical lights could be dimmed for reducing power consumption when daylight is available. Daylighting and shading devices such as light shelves, light pipes, and mirror ducts have potential for active control. Similarly, active technologies could be used for improving the performance of air conditioning and ventilation systems. Examples involving pre-cooling of intake air, energy recovery, thermal storage, and dehumidification are discussed. In addition to improving energy efficiency, comfort, convenience, safety, and security of buildings could be improved through automation. Finally, high level of automation in the management of buildings is achieved using facilities management systems integrated with building systems.

References

Agapaki, E. and Brilakis, I. (2021). CLOI: An Automated Benchmark Framework for Generating Geometric Digital Twins of Industrial Facilities, *Journal of Construction Engineering and Management*, 147 (11), art. no. 04021145.

Chen, Y., Raphael, B., and Sekhar, C.S. (2012). Individual Control of a Personalized Ventilation System Integrated with an Ambient Mixing Ventilation System, *HVAC&R*, 18 (6), pp. 1136–1152.

Crawley, D.B., Pedersen, C.O., Lawrie, L.K., and Winkelmann, F.C. (2000). Energy Plus: Energy Simulation Program, *ASHRAE Journal*, 42 (4), pp. 49–56.

Kasamsetty, S. and Raphael, B. (2018). Performance Evaluation of a High-Influx, Bubble Dehumidifier, *Energy and Buildings*, 173, pp. 291–301. https://doi.org/10.1016/j.enbuild.2018.05.047.

Omar, H., Mahdjoubi, L., and Kheder, G. (2018). Towards an Automated Photogrammetry-Based Approach for Monitoring and Controlling Construction Site Activities, *Computers in Industry*, 98, pp. 172–182.

Pantelic, J., Raphael, B., and Tham, K.W. (2012). A Preference Driven Multi-Criteria Optimization Tool for HVAC Design and Operation, *Energy and Buildings*, 55, pp. 118–126.

Perez-Perez, Y., Golparvar-Fard, M., and El-Rayes, K. (2021). Scan2BIM-NET: Deep Learning Method for Segmentation of Point Clouds for Scan-to-BIM, *Journal of Construction Engineering and Management*, 147 (9), art. no. 04021107.

Raphael, B. (2011). Active Control of Daylighting Features in Buildings, *Computer-Aided Civil And Infrastructure Engineering*, 26 (5), pp. 393–405.

Ward, L.G. and Shakespeare, R. (1998). *Rendering with Radiance: The Art and Science of Lighting Visualization*, San Francisco, CA: Morgan Kaufmann.

Warrier, G.A. and Raphael, B. (2017, 1 April). Performance Evaluation of Light Shelves, *Energy and Buildings*, 140, pp. 19–27.

5 Application

Construction automation

5.1 Introduction

> Once, this author was giving a demonstration of a construction automation system to a group of people from the industry. He said, "My dream is to have a machine into which you input a detailed 3D model of the building to be constructed and you press a button, and the building comes up automatically". One of the participants said, "That is not good – you should not have to press a button, you should just think, and the building should come up automatically". Today, researchers have demonstrated that we can build machines that can read the human mind, by reading electromagnetic signals passing through the brain and using pattern recognition to decipher the thoughts. So, the fantasy to develop machines that can build structures automatically by reading human thoughts may not be too far-fetched!

The concept of applying automation to construction was introduced in Chapter 1 (Section 1.3). This chapter will discuss the implementation details of this concept. A number of examples were discussed in the previous chapters. This chapter will attempt to follow a systematic approach to answer the questions: What can be automated? How can they be automated?

In industrial automation, there is a well-known methodology called the USA approach – Understand, Simplify, Automate. The first step is to understand the current processes, and carefully define all the components. Next, an attempt is made to simplify the current processes. Finally, the potential to automate is evaluated, and if the predicted benefits are promising, the automated solution is implemented. While this methodology can be applied to construction as well, there are difficulties because unlike in a factory, there are not many repetitive activities on a construction site. In construction activities, there are many components of processes, and these are not always well defined. Many activities are performed in an ad-hoc manner on site without prior planning because unanticipated conditions are encountered. Therefore, systematic application of the USA methodology is not possible. However, a modified version of this methodology is presented in the next section that helps to structure the discussion on what can or need to be automated.

In factories, the main activities can be grouped into the following categories:

- Movement of parts (transporting components from one station to the next)
- Processing (cutting, finishing, etc.)
- Assembly (connecting parts)
- Inspections (checking the quality of components)

DOI: 10.1201/9781003165620-6

On a construction site, similar activities are performed, but the activities are more complex and there are many possible variations of the basic activities depending on the context. Several researchers have tried to classify construction activities into basic types. One classification is given by Everett and Slocum (1994). The basic tasks listed by him are:

- Connect (e.g., screw, nail, bolt, weld)
- Cover (e.g., fix sheets)
- Cut (e.g., drill hole)
- Dig (e.g., excavate trench)
- Finish (e.g., grind marble floors)
- Inspect (e.g., verify alignment)
- Measure (e.g., calculate dimensions)
- Place (e.g., tiling)
- Plan (e.g., formulate work sequence)
- Position (e.g., erect steel beam)
- Spray (e.g., paint)
- Spread (e.g. pour concrete)

These tasks are composite activities that can be broken down into smaller parts having several basic movements. If we are able to identify the basic movements and define their characteristics precisely, we can design machines for automating these tasks.

This chapter will follow a case-study approach to discuss the possibilities for automation. Cases of activities on construction sites will be used to discuss the potential for automation. No attempt is made to prepare an exhaustive list of activities or to cover all possible automation solutions. The books by Bock and Linner (2016a, 2016b) have a fairly comprehensive list of existing automation systems. A few selected activities are taken to illustrate how to approach the problem of automation decision-making. The following cases of construction activities are considered in this chapter:

1. Surveying
2. Excavation
3. Transporting materials vertically (vertical delivery system)
4. Transporting materials horizontally (horizontal delivery system)
5. Placing material at precise locations
6. Pouring material
7. Cutting and removing material (processing parts)
8. Alignment and adjustments of parts
9. Connecting structural elements and parts (assembly)
10. Painting
11. Plastering
12. Progress monitoring (inspections)

5.2 Decision-making on construction automation

A structured approach to decide what automation solutions could be adopted for a construction activity is presented in this section. This involves understanding the current processes, simplifying the processes, and evaluating the possibilities for higher levels of automation. A structured analysis is carried out by answering the following questions.

a How is the activity commonly performed without automation?
b Can the process be defined precisely without any ambiguity? What are the inputs and the outputs? Can the activity be decomposed into standard subactivities? Can the activities be simulated on a computer?
c What are the movements required if the same activity is done by a machine (definition of kinematics)?
d What are the forces to be applied (analysis of dynamics)?
e What hardware is needed for automation?
f What software is needed?
g What are the technical challenges?
h What are the barriers?
i How can the benefits and effectiveness of automation be quantified?

The approach starts with understanding current manual practices and clearly defining the process (questions a, b). After that, we try to bring out the requirements for mechanization (questions c, d). Then we go deeper into hardware and software requirements for achieving the next level of automation (questions e, f). Finally, we think about the technical and economic feasibility (questions g, h, and i).

In general, identifying the automation hardware should start with a literature review or technology survey. Commercially available components should be identified to cater to the primary requirements of kinematics and dynamics of the task. These components might have to assembled in novel ways to develop cost-effective solutions. For many activities, hardware for the next generation of automated techniques is already available. Some examples of these are discussed in the following sections. For other cases, we consider hypothetical solutions that appear to be feasible from the current state of technological development.

For most automation solutions, understandably, cost turns out to be a barrier. Requirement for highly skilled labor is also a common barrier. Resistance of the society to replace human workers with machines is another common theme.

For evaluating the effectiveness of automation, common methods include estimation of the savings in time and cost. These can be compared with traditional methods if enough data are available. Data can be collected using the process of activity-based work sampling. This involves visiting a site, making observations of actual activities that are performed on site, and recording the time taken for a detailed breakdown of the activities. The number of observations to be made is based on established statistical procedures using an estimate of the required confidence levels. After collecting basic data, the process could be simulated on a computer to calculate the time taken (Krishnamoorthi and Raphael, 2021). Discrete Event Simulation (DES) is a promising approach to model complex construction processes (Krishnamoorthi and Raphael, 2018).

The approach of structured analysis is illustrated using a few case studies in the following sections. It is not claimed that the answers to all the questions are provided conclusively. Some pointers to possible answers are provided. Readers should explore other concepts and ideas, developing these into detailed designs that might be implemented with commercially available hardware components. This author has used some of these examples for brainstorming exercises in automation workshops conducted in colleges as well as companies. In most examples presented later, the question b is omitted. This is because the current processes are not well defined. There are many variations of possible options and there are many details that are specific to each option. It will be an interesting exercise to take specific solutions and work out the details of the complete process model. Developing simulation models and visualizing them on a computer will also be useful. Accurate visualization on a computer is a good indicator that the process is well-defined.

5.3 Examples of construction activities that might be automated

5.3.1 Surveying

The purpose of land surveying on a construction site is to determine the relative elevations of points on the site. In the past, chains and tapes were used for measuring horizontal distances, and theodolites were used to measure vertical levels (elevation differences) as well as angles. Today, total stations are commonly used for surveying. These are more sophisticated devices using Electronic Distance Measurement (EDM).

Exercise 5.1

Watch videos that show the use of total station on the internet. Make a list of the manual operations that are needed for measuring the coordinates of points on a site.

Structured analysis

HOW IS THE ACTIVITY COMMONLY PERFORMED WITHOUT AUTOMATION?

Today, it is impossible to think of doing surveying without any automation. Already, total stations have high levels of automation. Laser devices are used to measure distances electronically; angles are also measured using accurate optical devices. Distance measurement is typically done by emitting a beam of light, which is reflected back using a prism reflector. The time of flight of the reflected light or the difference in phase angles of light waves is used to compute the distance. It is also possible to measure distances without the use of a reflector, by aiming at a hard surface; however, errors are likely to be high with this.

WHAT ARE THE MOVEMENTS REQUIRED (KINEMATICS)?

The total station is mounted at a reference location and a reflector is fixed at the point where the coordinates need to be measured. The total station is leveled by adjusting the tripod to make sure that the base is perfectly horizontal. Spirit levels could be used to make sure that the base is horizontal; electronic angle measuring devices (inclinometers) could also be used. The laser is pointed to the reflector, and the correct point is selected by checking through a view finder. After these operations are done, measurement is as simple as pressing a button. Many measurements are typically recorded on the device and transferred to a computer using appropriate means.

WHAT ARE THE FORCES TO BE APPLIED (DYNAMICS)?

Since the measurement is mostly electronic, large forces need not be applied. Nevertheless, actuators are needed for automating different movements. Leveling, rotation of the view finder, etc., are operations that require low levels of forces.

AUTOMATION SOLUTION: WHAT HARDWARE IS NEEDED FOR AUTOMATION?

While sophisticated total station devices have significant levels of automation, surveying still takes considerable amount of time. For the next level of automation, radical changes in the procedure are needed. A promising technology is laser scanning using drones. An unmanned aerial vehicle flies over the terrain, sends laser pulses to the ground, and measures coordinates. This laser scanning process results in a large "point cloud" which contains the x, y, z coordinates of many points. A model of the terrain can be reconstructed using this point cloud data.

WHAT SOFTWARE IS NEEDED?

Algorithms are available for digitally reconstructing the surface that is scanned. Geometric models based on mesh representations are easily generated. However, more sophisticated processing of the data might require advanced techniques using machine learning. For example, recognizing objects and features in the terrain is not easy with conventional techniques. In many cases, an entire surface may not be scanned from a single point since there may not be line of sight. Multiple scans are needed from different viewpoints. Merging data from multiple viewpoints and generating a consistent and accurate geometric model requires complex algorithms.

WHAT ARE THE TECHNICAL CHALLENGES?

Managing large volumes of data and filtering out noise in data are challenges. For detecting objects and features in the terrain, complex algorithms are needed. Engineers with knowledge in advanced computational techniques are needed for these tasks. If drones are used, autonomous navigation avoiding obstacles, automatically determining the flight path with minimum input from user, etc. are tasks that still require lot of research.

WHAT ARE THE BARRIERS?

The use of drones is regulated in many places. They cannot be deployed everywhere. Current technological immaturity is also a barrier.

HOW TO QUANTIFY THE BENEFITS AND EFFECTIVENESS OF AUTOMATION?

To determine the effectiveness of automation, reliable data are needed under diverse conditions. The time taken for setting up the drone, completing the scanning, and processing the data needs to be measured. This can then be compared with conventional surveying methods to determine the effectiveness of automation.

5.3.2 Excavation

Excavation is a task that is widely discussed in publications on construction automation. Excavation is an essential activity in many construction projects; in mining it is the most important activity. Many types of heavy equipment have been developed for completing this task faster.

Exercise 5.2

Study images and videos of excavators on the internet. Identify the parts of an excavator.

Structured analysis

HOW IS THE ACTIVITY COMMONLY PERFORMED WITHOUT AUTOMATION?

Excavators are vehicles with hydraulic booms and buckets, commonly used for large excavations. Skilled and experienced operators use joysticks to operate the hydraulic controllers of the boom and bucket.

WHAT ARE THE MOVEMENTS REQUIRED (KINEMATICS)?

The vehicle, a truck, is driven to the location on site. It is kept still at the position where earth should be cut and removed. The operator sits inside a cabin. The entire cabin along with the arm and the bucket can be rotated 360° so that soil can be removed from anywhere around the machine. The boom is lifted using a hydraulic cylinder. There are separate hydraulic cylinders for rotating the arm and lifting the bucket. All these are controlled using joysticks, which are manually operated by the human operator.

First, the bucket is lowered to cut into the soil. When the bucket is full, it is lifted, and the soil is dumped into a loading truck. This requires turning the machine and bringing the bucket above a waiting truck. When the truck is full, it is driven away to the location where the material is dumped.

WHAT ARE THE FORCES TO BE APPLIED (DYNAMICS)?

Large forces are needed for cutting into soil, lifting the soil, etc. Hydraulic actuators with high power requirements are used for these activities.

AUTOMATION SOLUTION: WHAT HARDWARE IS NEEDED FOR AUTOMATION?

Many manual operations involved in excavation might be automated. Since an excavator is a heavy equipment, it may not be easy to fabricate custom-designed machines. Our options are limited to those provided by equipment manufacturers. Some of the manufacturers are already producing versions with significant automation. Some of the features that are already there or likely to be introduced are briefly discussed here.

1 Use of a geometric model of the trench to be excavated: The model is uploaded into the onboard computer. The machine is first positioned at a precise location that is referenced in the model. When the bucket of the excavator hits the ground, the computer calculates the depth to which digging must be carried out. The computer can take over from here, and complete the task with minimum intervention from the user. The exact shape of the trench is known to the computer, and the necessary operations of the tools are carried out automatically.
2 Instead of joysticks and pedals for directly manipulating the tools, the computer software receives the data, becoming the input device, and interacts with objects on a screen onboard the vehicle. The boom and the bucket are virtually controlled on the screen.

Virtual reality and augmented reality could be used for enhancing the experience of human-machine interaction.

3 Sensors that measure various aspects of the operations are used to determine safe operations. For example, the pressure applied by the bucket on the ground, soil conditions, the present slope of the ground, and environmental parameters are used for the optimal operations.
4 The optimal path could be computed automatically given the geometry of the terrain. The time could be estimated and adjusted according to the conditions of the soil.

WHAT SOFTWARE IS NEEDED?

From this description of hardware, it is clear that machine learning and artificial intelligence play a crucial role in the development of a fully automatic excavator.

WHAT ARE THE TECHNICAL CHALLENGES?

The challenges relate to both hardware and software. Since large forces are involved, it is difficult to develop prototypes and test them for performance; equipment development can be performed only by large manufacturers. Software: Robust machine-learning techniques are needed; for example, reinforcement learning might be relevant. These are still active research areas.

WHAT ARE THE BARRIERS?

As in most automation technology, cost and technical skill of workers are likely to be the major barriers in widespread deployment.

HOW TO QUANTIFY THE BENEFITS AND EFFECTIVENESS OF AUTOMATION?

Since fully automated excavators are rare and not much data are available, it is difficult to judge now. The robustness and reliability of such machines under harsh conditions need to be evaluated. Once reliable data are available, processes might be simulated to quantify the benefits.

Exercise 5.3 Group discussion

a Analyze the practical feasibility of using automated excavators that use digital models for the preparation of the site.
b What other solutions might be possible for improving the automation of excavation operations?

5.3.3 Transporting materials vertically (vertical delivery system)

In many countries, especially for small construction in developing countries, material is transported manually. Laborers carry the load and climb stairs or lifts to the required locations. Hoists and pulleys are also used to lift construction materials to upper floors of buildings. These are usually manually operated; sometimes motorized hoists are used. For large construction, tower cranes are the preferred choice. However, this requires careful planning and

Figure 5.1 A mobile crane lifting a heavy pipe

involves large expenses to set up the cranes. Mobile cranes mounted on trucks are used to lift materials to small heights. Since these technologies have become common and affordable, manual lifting of materials is not discussed further in this section. A mobile crane is taken as an example of the current "manual" process. A picture of a mobile crane lifting a heavy pipe is shown in Figure 5.1.

Exercise 5.4

Study images and videos of mobile boom cranes on the internet. Find out how many degrees of freedom are present in a typical boom crane.

Structured analysis

HOW IS THE ACTIVITY COMMONLY PERFORMED WITHOUT AUTOMATION?

The operation of a mobile crane is explained here for illustration. (These activities are best understood by watching videos). The vehicle is placed in position at the location from where material should be lifted. A lever is pulled to release the legs of the stabilizer (outrigger) and the legs are extended to all the sides. The legs are lowered to touch the ground. Then the vehicle is slowly lifted such that the entire load is supported on the legs. The legs are adjusted such that the vehicle is leveled and in a stable position.

The operator sits inside the cabin that contains all the controls. There are several levers for operating the crane: a) to release and retract the hook for holding the load; b) to lift and lower the crane; c) to control the telescopic extension of the boom; d) to rotate the crane.

Once the crane is ready, the crane is rotated, the boom is extended, the hook is lowered and brought to the position where the object to be lifted is kept. The hook is connected to the weight and the weight is lifted. The crane is rotated and the hook is retracted or extended to reach the position where the weight is to be placed. Then the hook is disconnected from the weight.

WHAT ARE THE MOVEMENTS REQUIRED (KINEMATICS)?

Aside from the operations needed to set up the outrigger, movements are needed in the x, y, and z directions to bring the hook to the source or the destination. The operator uses levers for these operations. The other major operation is carefully balancing the load and connecting the load to the hook.

WHAT ARE THE FORCES TO BE APPLIED (DYNAMICS)?

Large forces are needed for the lifting operation. Hydraulic systems are typically used for this.

AUTOMATION SOLUTION: WHAT HARDWARE IS NEEDED FOR AUTOMATION?

The mobile cranes are semi-automated equipment. Computer control is needed for the operation of the telescopic boom. Complex control is needed for other operations such as balancing the outriggers and ensuring safety at every stage. However, it is still a manually operated piece of equipment. Careful maneuvering is needed to accurately position the hook at the source and destination. The entire process is heavily dependent on skilled operators. Wrong operations could cause collisions and serious accidents.

As we move to the next level of automation, safety could be improved using sensors and cameras. Accidents could be avoided by automated analysis of images and sensor data; these could be used for taking appropriate control actions in dangerous situations. For example, if a worker or vehicle approaches the crane beyond safe distance, warning could be given, and the operations could be stopped.

For fully automated material handling, construction processes must be changed radically. One-off lifting and placing of objects should not happen. Precise modeling of the requirements to move material on site should be made on a computer. The movements should be optimized such that multiple elements that are to be lifted to the same location should be packed appropriately in containers which are properly tagged and lifted together. The lifting system should resemble the automated movement of containers in a warehouse. Manually connecting weights to hooks, positioning by trial and error, etc. can be eliminated.

WHAT SOFTWARE IS NEEDED?

Computer-vision techniques will be useful for improving safety. Machine learning could be used for analyzing other types of sensor data for safe operations. Optimization and path planning are crucial for automated movement of materials.

WHAT ARE THE TECHNICAL CHALLENGES?

Computer vision and image processing are still research areas. Robust and reliable techniques are still not available in complex environments such as construction sites. Detailed modeling of all the logistics on a construction site is also challenging. Efficient methodologies have yet to be developed for these.

WHAT ARE THE BARRIERS?

There might be legal and ethical issues related to completely autonomous material handling processes (this might be a common factor in most automated tasks).

HOW TO QUANTIFY THE BENEFITS AND EFFECTIVENESS OF AUTOMATION?

Material handling processes can be simulated to find out the time taken for various options. Time and cost are important parameters in determining the effectiveness of automation.

Exercise 5.5 Group discussion

a Analyze the practical feasibility of digitally modeling every site operation prior to starting the construction. Find out if there are additional challenges in implementing the proposal.
b Discuss other potential solutions for improving the vertical delivery systems.

5.3.4 Transporting materials horizontally (horizontal delivery system)

In this section, we consider the movement of objects within a floor of a building, that is, at the same level of construction. Horizontal material transport is heavily dependent on manual labor in most countries. Relatively lightweight material such as tiles, bricks, and mortar are manually moved for short distances. For moving these in bulk, mechanical systems such as wheelbarrows and trolleys are used, but these require operations such as loading, unloading, searching, sorting, etc. Since these are not considered major activities in a project plan, people tend to overlook the time required for these movements. These are small repetitive movements that could have a ripple effect on the overall time taken by larger activities. Hence, it is important to estimate the actual time taken for these activities.

In factories, conveyor belts are used to move material such that workers need to make minimum movements for completing their work. Instead of making workers move, material is made to move to the locations of workers. This is possible in prefabrication factories, but it is difficult on construction sites because of the dynamically changing locations of work and materials.

Manually operated forklifts help to alleviate the use of physical labor for material movement. It could also improve the cycle time of activities if material is properly packed and organized, and the entire movement is properly planned. Tagging the packages using RFID or other schemes help in reducing the time required to searching the right location to deliver the material and to locate the right material for a task.

Automated guided vehicles have been used in warehouses. However, construction sites are different; they are very chaotic and cluttered. It is dusty and the environment is harsh. If automated guided vehicles are designed to work in such harsh conditions, that could provide a solution to the horizontal transport problem.

Exercise 5.6

Watch videos of the use of automated guided vehicles in warehouses. What are the differences in the working conditions between warehouses and construction sites?

HOW IS THE ACTIVITY COMMONLY PERFORMED WITHOUT AUTOMATION?

As mentioned earlier, there are many different tasks that involve use of human arms and limbs. Lifting, loading, pushing, unloading, arranging, etc. are common activities.

WHAT ARE THE MOVEMENTS REQUIRED (KINEMATICS)?

Gripping, lifting, horizontal movements, placing, un-gripping, are the required movements. Unlike robots in factories, material is moved over larger distances. Hence typical ranges in the x and y directions of stationary industrial robots will not be sufficient.

WHAT ARE THE FORCES TO BE APPLIED (DYNAMICS)?

Lifting requires forces that are determined based on the weight of the object. Horizontal transport could also require quite significant forces to overcome friction during the movement.

AUTOMATION SOLUTION: WHAT HARDWARE IS NEEDED FOR AUTOMATION?

Fully automated solutions involve pre-programmed picking and placing of boxes or containers. It may not be economical to install gantry systems or cranes on every floor. However, motorized versions of forklifts that resemble automated guided vehicles are feasible. Futuristic solutions could also be imagined, such as wheeled robots that follow hand gestures and voice commands to move material; these do not require the coordinates to be pre-programmed, and there will be more flexibility in carrying out the work. These "delivery agents" could be programmed to collect packets from the vertical delivery system, automatically determine the delivery locations from their tags, determine the shortest path to be taken considering the actual site constraints, and deliver the material at the precise location without any human intervention.

WHAT SOFTWARE IS NEEDED?

Robotic solutions need computer vision, object recognition, and many other features which are promised by artificial intelligence.

WHAT ARE THE TECHNICAL CHALLENGES?

Designing electrical and mechanical systems for the harsh conditions of a construction site is a challenge. Also, wireless communication tends to be unreliable because of the obstructions and heavy metallic components on site.

WHAT ARE THE BARRIERS?

Technical skills needed for operating the machines, hesitance of society to replace humans with machines, etc. are barriers.

HOW TO QUANTIFY THE BENEFITS AND EFFECTIVENESS OF AUTOMATION?

The effectiveness of automation might be evaluated through simulations. Work sampling methodology that has been adopted in lean construction is also useful for measuring actual performance improvements on site.

Exercise 5.7 Group discussion

a Analyze the practical feasibility of automated guided vehicles on construction sites.
b Discuss what other solutions might be possible for improving the horizontal delivery systems.

5.3.5 Placing material at precise locations

Many construction activities require placing material precisely at specific locations. When we discussed vertical and horizontal transport systems, we did not place much emphasis on precision; material could be placed roughly around the task location. In this section, we discuss activities that require the precision of less than a couple of millimeters. An example is bricklaying. Every brick must be vertically and horizontally aligned – otherwise, there will be large errors over time. Precise vertical alignment is important for the stability of the structure. Even though, there are many other activities that require precision; we restrict the discussion in this section to bricklaying for constructing masonry walls.

Exercise 5.8

Watch videos of bricklaying robots on the internet. Comment about the complexity of the machine and its cost implications.

Structured analysis

HOW IS THE ACTIVITY COMMONLY PERFORMED WITHOUT AUTOMATION?

A mason picks up the brick and places it on the wall under construction after putting down a layer of mortar. A thread, which is tied from one end of the wall to the other, is used for checking the horizontal alignment. A plumb bob is used to ensure that the wall is strictly vertical. The brick is hit with a trowel or hammer to adjust its position if the alignment is not perfect. Usually, there is a helper who is an unskilled laborer to help the skilled mason with tasks such as lifting the brick, placing it within close reach of the mason, etc.

WHAT ARE THE MOVEMENTS REQUIRED (KINEMATICS)?

For placing the brick: vertical movement (z) is needed for lifting the brick from the stacked location to the actual height where it is placed. Depending on the shape of the wall, the brick should be moved in the horizontal directions (x, y) to the exact position where it should be placed. Fine adjustments in the x, y, and z directions are needed for vertical and horizontal alignment.

For placing mortar: Since this is not a solid object, standard picking and placing solutions are not applicable. Pumping the mortar is feasible, but it requires pumping small volumes in precise quantities. Then the placed mortar should be pressed to create a flat surface.

WHAT ARE THE FORCES TO BE APPLIED (DYNAMICS)?

Force equivalent to the weight of the brick needs to be applied for vertical lifting. Small forces are needed to move the brick horizontally for alignment.

AUTOMATION SOLUTION: WHAT HARDWARE IS NEEDED FOR AUTOMATION?

Bricklaying robots have been developed by a few companies. A common design consists of a robotic arm with a gripper that picks up a brick from a stack, lifts it to the location where it is to be placed, rotates it to the right orientation and places it. A standard 6-DOF industrial robot is sufficient for these operations. However, the design can be simplified since the required movements are not as complex as that in a manufacturing plant. Simple x, y, z movements are required. The robots are usually mounted on wheels or rails, since they must travel longer distances than the horizontal range of a standard robotic arm that is mounted at a stationary location. In some machines, mortar has to be manually placed on the bricks before they are loaded. Automatically pouring mortar, filling it inside the cavities, etc. appear to be complex operations.

More sophisticated robots might have sensors for visual inspection of the completed work, evaluate the quality of work, and take rectifying actions if there are defects. Computer vision might be needed to detect obstructions and avoid dangerous situations. Other sensors might be used to determine whether the bricks have been correctly loaded, they are of the right size, etc.

WHAT SOFTWARE IS NEEDED?

The plan for placing the bricks should be computed automatically from a geometric model of the wall. Software for the control of the robot should account for special conditions such as edges of walls and the top layer when full bricks do not fit inside the available space. Computer vision and sensor data processing might require machine learning techniques.

WHAT ARE THE TECHNICAL CHALLENGES?

There are many possible situations on the construction site which cannot be anticipated and planned. Developing a robust system that works well in all these situations is a challenge.

WHAT ARE THE BARRIERS?

Robotic solutions tend to be expensive. Careful calculations are needed to establish return on investment. Accurate planning is needed to ensure that the equipment is fully utilized all the time so that investment is recovered quickly. In practice, many contractors do not have such skills to make effective use of expensive equipment.

HOW TO QUANTIFY THE BENEFITS AND EFFECTIVENESS OF AUTOMATION?

Productivity should not be the only criterion to determine the effectiveness of automation. Fatigue and health issues of workers doing repetitive activities and related issues should also be considered in the analysis. Current manual construction of masonry walls is in a way constrained by the capabilities of workers to lift and manipulate the bricks. With robotic masonry, large size blocks and prefabricated panels could be considered. These might provide better efficiency and productivity.

Exercise 5.9 Group discussion

What other solutions are attractive for non-load-bearing wall construction? Are these likely to replace brick masonry?

5.3.6 Pouring material

Materials having different properties need to be poured during construction. This includes concrete, mortar, grout, and water. Since the flow properties are different, separate equipment are needed for each type of material. In this section, the example of pouring concrete will be taken for illustration.

Exercise 5.10

Watch videos of concrete boom pumps in operation. List the activities done by the workers during the operation.

HOW IS THE ACTIVITY COMMONLY PERFORMED WITHOUT AUTOMATION?

A truck-mounted concrete boom pump is commonly used in large construction. This equipment can pump concrete to reasonably large heights. Even though several automation features are present, most of the operations are manual. Some operations are similar to a boom crane: outriggers are deployed, and the truck is stabilized. Then, concrete is poured into the hopper, typically from a mixer or a ready-mix truck. The boom is maneuvered using a remote control or joystick to position the hose over the location where concrete needs to be poured.

WHAT ARE THE MOVEMENTS REQUIRED (KINEMATICS)?

The boom needs to be controlled to position the hose at the location where the concrete needs to be discharged. This requires movement in the x, y, and z directions. In addition, technicians need to operate the levers for functions such as stabilizer setup, starting, and stopping of pumps, etc.

WHAT ARE THE FORCES TO BE APPLIED (DYNAMICS)?

Pumping concrete requires large pressure. Heavy pumps and hydraulic cylinders are used for this.

AUTOMATION SOLUTION: WHAT HARDWARE IS NEEDED FOR AUTOMATION?

Even though concrete pumping is largely an automated operation, considerable amount of time is required for preparing the formwork, scaffolding, rebar cage, etc. These are mostly manual operations. Some activities such as preparing the rebar cages can be shifted to on-site factories or off-site factories and can be automated. However, formwork must be assembled on site. The quality of formwork affects the quality of the final product. In particular, the formwork needs to aligned precisely, vertically and horizontally. Instead of automating these tasks on site, it looks more promising to shift the entire concreting process to factories. Prefabrication, modular construction, and 3D printing are emerging technologies. Prefabricated large modules can be assembled on site with higher levels of automation.

Concrete 3D printers have already been demonstrated by many researchers. Concrete 3D printers deposit specially designed mixes of concrete, layer by layer, to achieve the required shape of the element. Robotic arms could be integrated with concrete 3D printers so that reinforcement bars and cages can be inserted into the 3D printed objects. Structural elements and larger modules that have the required strength can be fabricated using this approach. These elements could be lifted and assembled on site.

WHAT SOFTWARE IS NEEDED?

3D printers take geometric models to correctly print the shape. Software to process the geometry and generate the path for the 3D printer is required. Optimization of path and operations of the robotic system might be needed. If printers are equipped with sensors to provide feedback on the quality of the printed elements, defects could be corrected automatically.

WHAT ARE THE TECHNICAL CHALLENGES?

Concrete 3D printing is an emerging area. There are several technical challenges including developing mixes with the required flow properties, integration of robotics for placing reinforcements, and controlling unwanted deformation of the deposited concrete during the printing process.

WHAT ARE THE BARRIERS?

Like many automation technologies, cost is a barrier. Until mass production is possible, concrete 3D printers will be expensive. Many technological demonstrations are needed to convince people about the technical viability and economic advantages of concrete 3D printing.

HOW TO QUANTIFY THE BENEFITS AND EFFECTIVENESS OF AUTOMATION?

The greatest advantage of concrete 3D printing is the elimination of requirement of formwork, which reduces the time, cost, and the waste of material. An accurate economic analysis of the process taking all these factors into account will reveal how effective is automation with this approach.

Exercise 5.11 Group discussion

WHAT ARE THE PRACTICAL SITUATIONS WHERE CONCRETE 3D PRINTING IS BENEFICIAL AND COST-EFFECTIVE?

5.3.7 Cutting and removing material

In industrial manufacturing, subtractive manufacturing is a commonly used technique. A large object is taken and parts are removed from it to create the required shape. In construction, subtractive manufacturing is not so common. But we face many situations where we need to remove material. We need to drill concrete slabs and walls for inserting pipes and ducts; cut lifting hooks after completing precast elements; cut tiles and panels to get the correct shape that fits within available space, etc. A picture of a masonry wall that was cut for a ventilation duct is shown in Figure 5.2.

HOW IS THE ACTIVITY COMMONLY PERFORMED WITHOUT AUTOMATION?

For cutting rebars, cutting machines are available. Drills are available for making holes in concrete. These are usually manually operated. First, precise locations are determined through appropriate measurement and marking techniques. Then the tool is used for cutting

Figure 5.2 A masonry wall that was cut for a ventilation duct

and removing material. Machines are also available for cutting large chunks of concrete, such as parts of slabs and beams.

WHAT ARE THE MOVEMENTS REQUIRED (KINEMATICS)?

Typically, two-dimensional movements are sufficient after the tool is inserted into the workpiece. Cutting is usually performed along a 2D path.

WHAT ARE THE FORCES TO BE APPLIED (DYNAMICS)?

Reasonably large forces are needed for drilling and cutting concrete and metal pieces.

AUTOMATION SOLUTION: WHAT HARDWARE IS NEEDED FOR AUTOMATION?

In construction, cutting and removing material is usually a one-off activity. Other than using mechanical tools, automation is difficult. It might be more feasible to eliminate the need for cutting through proper planning and changing the construction process. The cut made in the wall in Figure 5.2 could have been avoided with better planning. However, there are other activities where cutting is more efficient than constructing the actual shape. For example, CNC (computer numerial control) machines could be used efficiently for tasks such as cutting timber for formwork and metals for steel fabrication.

WHAT SOFTWARE IS NEEDED?

Digital modeling software will help in planning all the requirements to place ducts and avoid having to drill holes and cut material. Optimization of processes will be needed.

WHAT ARE THE TECHNICAL CHALLENGES?

The main technical challenge is managing information and anticipating all the requirements from various stakeholders in construction.

WHAT ARE THE BARRIERS?

Lack of skills in using advanced planning tools might be the biggest barrier.

HOW TO QUANTIFY THE BENEFITS AND EFFECTIVENESS OF AUTOMATION?

Effectiveness of the proposed solution can be evaluated by taking into account the time taken for cutting and the material wasted in the process.

Exercise 5.12 Group discussion

ARE THERE SITUATIONS IN WHICH CUTTING AND REMOVING CANNOT BE ELIMINATED DURING CONSTRUCTION?

5.3.8 Alignment and adjustments of parts

As discussed previously in several examples, alignment and adjustments are needed for many activities. Bricks need to be aligned to maintain verticality of walls, and formwork needs to be aligned to match the precise positions on the drawings.

HOW IS THE ACTIVITY COMMONLY PERFORMED WITHOUT AUTOMATION?

The common procedure is marking the positions at both ends of the wall and then tying a thread between the two points. Threads are pulled and kept tight to visually check whether a straight line is maintained. Plumb bobs are used to check vertical alignment. Today, the use of laser devices for alignment is more common. A laser beam is used to check visually and ensure that the walls and other elements maintain a straight line according to the design.

WHAT ARE THE MOVEMENTS REQUIRED (KINEMATICS)?

If the laser alignment is to be done by machines, the reference point needs to be located, and the laser source needs to be moved to this position. Then the beam should be rotated to point in the direction of alignment.

WHAT ARE THE FORCES TO BE APPLIED (DYNAMICS)?

Forces are not critical for this task. Nominal forces are required for the operations.

AUTOMATION SOLUTION: WHAT HARDWARE IS NEEDED FOR AUTOMATION?

Automation in alignment could be improved using laser scanners. Accurate 3D geometry of the object under construction could be obtained through laser scanning. Any deviation in any direction can be detected immediately. Corrective actions to maintain the alignment could be taken. Alternatively, an automated picking and placing system could make use of laser scanning data to position the material correctly without the need for adjustments.

WHAT SOFTWARE IS NEEDED?

Software to process point cloud data from laser scanner is required. A robust monitoring system that makes use of this data in real time will be needed.

WHAT ARE THE TECHNICAL CHALLENGES?

Processing laser scanning data are computationally expensive. Mapping point cloud data to the design geometry (registration) involves several challenges.

WHAT ARE THE BARRIERS?

The cost of laser scanners might be the most critical barrier. Until software becomes more user-friendly, it might also be time-consuming to set up and process the data.

HOW TO QUANTIFY THE BENEFITS AND EFFECTIVENESS OF AUTOMATION?

To answer this question, we need reliable data. If data are collected through work sampling, we can compare the time taken for the current manual processes and that for laser scanning.

Exercise 5.13 Group discussion

a Discuss other possibilities for the automated alignment of formwork for cast-in-situ walls and columns.
b How do you reduce the cost of laser scanning?

5.3.9 Connecting structural elements and parts

Connecting structural elements is critical in prefabricated construction. Whether the elements are made of concrete or steel, the connections must be carefully designed such that the forces are effectively transmitted at the joints. In steel structures, connections are usually bolted or welded. As an example, only welded steel connections are discussed here.

Exercise 5.14

Watch videos of welding robots used in factories. List the challenges in using these on construction sites.

HOW IS THE ACTIVITY COMMONLY PERFORMED WITHOUT AUTOMATION?

Steel parts are carefully aligned and temporarily supported on site. A qualified welder uses a welding machine to join the parts according to the specifications.

WHAT ARE THE MOVEMENTS REQUIRED IF THE SAME ACTIVITY IS DONE BY A MACHINE (KINEMATICS)?

Alignment of parts require x, y, and z movements and rotations of the parts. Welding is typically done along a straight line. This requires moving the welding machine according to a predefined path.

WHAT ARE THE FORCES TO BE APPLIED (DYNAMICS)?

Forces required for the welding operation are relatively small. Alignment requires large forces to move heavy structural elements.

AUTOMATION SOLUTION: WHAT HARDWARE IS NEEDED FOR AUTOMATION?

Automated welding machines have already been used in heavy construction. Welding robots are commercially available.

WHAT SOFTWARE IS NEEDED?

A digital model of the structure with low-level details of connections are needed. Software to use this model to control the robotic welder is required.

WHAT ARE THE TECHNICAL CHALLENGES?

Developing detailed geometric models of the connections is time-consuming. Simplifying the modeling tools to make the task easier for the operators requires more research in human computer interaction.

WHAT ARE THE BARRIERS?

Welding robots are expensive. Operating them might require higher level of technical skills from the workers.

HOW TO QUANTIFY THE BENEFITS AND EFFECTIVENESS OF AUTOMATION?

Apart from the time that is saved through automatic welding, enhanced safety should also be included in the analysis. Performing welding at heights and in difficult orientations involve quite unsafe conditions, which might be eliminated by the use of robots.

Exercise 5.15 Group discussion

a What other automated solutions might be possible for connecting steel parts?
b What are the challenges in making bolted connections using robots?

5.3.10 Painting

Painting is mostly done manually since building elements are large. In automobile factories, some parts are painted by dipping the entire part into paint. Spray painting is also automated in many factories. These are difficult to replicate on construction sites; however, there is potential in prefabricated parts.

Exercise 5.16

Watch videos of spray painting of automobile bodies in factories. Discuss why we are not adopting these solutions for precast wall panels.

HOW IS THE ACTIVITY COMMONLY PERFORMED WITHOUT AUTOMATION?

A human painter climbs on a supporting platform. A brush is dipped into the paint that is prepared and the paint is applied on the surface using the brush.

WHAT ARE THE MOVEMENTS REQUIRED IF THE SAME ACTIVITY IS DONE BY A MACHINE (KINEMATICS)?

In construction we have mostly horizontal or vertical surfaces that are flat. If machines are to do painting, we need mostly two-dimensional movements.

WHAT ARE THE FORCES TO BE APPLIED (DYNAMICS)?

Small forces are enough for the painting operations.

AUTOMATION SOLUTION: WHAT HARDWARE IS NEEDED FOR AUTOMATION?

Commonly available industrial robots might be used for painting. The end-effector could be equipped with a pump that sprays paint on the surface.

WHAT SOFTWARE IS NEEDED?

A digital model of the structure is needed. Software using this model to control the painting robot welder is required.

WHAT ARE THE TECHNICAL CHALLENGES?

Accurately controlling the robot in congested areas and corners requires robust path planning. Paint should be applied precisely at the required spots without spilling over to other parts.

WHAT ARE THE BARRIERS?

Like all robotic solutions, cost might be a significant barrier. Operating robots require higher levels of technical skills from the workers.

HOW TO QUANTIFY THE BENEFITS AND EFFECTIVENESS OF AUTOMATION?

Since large areas need to be painted, automated solutions are likely to be effective for this task.

Exercise 5.17 Group discussion

a Can we completely eliminate the need for painting of building surfaces?
b What other solutions might be possible for improving the automation of painting?
c Evaluate potential benefits in using a robotic system for painting the external walls of a tall building. The painting robot could be suspended using cables from the top of the building.

5.3.11 Plastering

Plastering seems to be similar to painting. However, this is a more difficult task than painting because of the properties of materials involved. Mortar cannot be pumped as easily as paint. There are solutions such as shotcrete. Shotcrete is a method of applying concrete by spraying concrete at high velocity on the surface. The concrete sticks to the surface and sets. This is currently mostly used for tunnels but could be used for plastering as well.

Exercise 5.18

Watch videos of shotcreting. Compile a list of projects where they have successfully used this technique outside tunneling.

HOW IS THE ACTIVITY COMMONLY PERFORMED WITHOUT AUTOMATION?

Mortar is picked using a trowel and pressed against the surface by the mason. Then the surface is given a smooth finish.

WHAT ARE THE MOVEMENTS REQUIRED IF THE SAME ACTIVITY IS DONE BY A MACHINE (KINEMATICS)?

Since plastering is typically done on flat surfaces, two-dimensional movements are sufficient for covering the entire area. If plaster is not pumped or sprayed, additional degrees of freedom are needed for placing the plaster precisely and providing a smooth finish.

WHAT ARE THE FORCES TO BE APPLIED (DYNAMICS)?

Pumping and spraying require quite large forces.

AUTOMATION SOLUTION: WHAT HARDWARE IS NEEDED FOR AUTOMATION?

Commercially available industrial robots might be used for plastering as well. The end-effector could be equipped with a pump that sprays plaster on the surface. More complex robotic designs might also be developed for mimicking human operations that results in a good quality finish for the surfaces. It should also be ensured that the plastered surface is not porous so that water does not get absorbed, making the wall damp during rains. The right amount of pressure needs to be applied to obtain the required properties of the plastered surface.

WHAT SOFTWARE IS NEEDED?

Sophisticated control software is needed, especially to take care of difficult areas such as corners and where there are electrical installations.

WHAT ARE THE TECHNICAL CHALLENGES?

Accurately controlling the robot in congested areas and corners requires robust path planning.

WHAT ARE THE BARRIERS?

Like all robotic solutions, cost might be a significant barrier. Operating robots require higher levels of technical skills from the workers.

HOW TO QUANTIFY THE BENEFITS AND EFFECTIVENESS OF AUTOMATION?

Since large areas need to be plastered, automated solutions are likely to be effective for this task. Work sampling or simulations might be used for quantification of the benefits.

Exercise 5.19 Group discussion

a Discuss the feasibility of developing a robotic system for plastering that mimics human operations.
b What movements are needed for a machine to provide a good finish at the corners of walls?

5.3.12 Progress monitoring

The conventional method of collecting data about the progress of construction is through daily, weekly, or monthly progress reports that are prepared manually. This information is used for releasing progress payments for competing milestones and for management control. For decision-making, the project manager requires an accurate model of the part of the structure that is completed The manual process of preparing progress reports is time-consuming. First, the completed activities completed must be identified on site and inspected. Even a simple building project involves thousands of activities and building elements. Data about the elements that are completed must be compiled into information about the percentage of work completed. Second, information is input into unstructured documents, and extracting relevant information automatically from these is difficult.

How is the activity commonly performed without automation?

A representative of the client or project manager visits the site with a checklist. The elements that are completed are noted down in the checklist. Records related to these elements are retrieved from a database or spreadsheet; these are updated based on the site visit. Then a report is prepared to be submitted to the client.

What are the movements required if the same activity is done by a machine (kinematics)?

Even though this is a data collection task, physical movement to the location of construction activity might be needed.

What are the forces to be applied (dynamics)?

Forces are not relevant for this task, since this is mostly a task related to gathering information.

Automation solution: what hardware is needed for automation?

Physical inspection may be replaced by virtual data collection. Cameras, laser scanners, and tracking devices could be installed on site. Data collected by these devices can be transmitted to a central database automatically.

In addition, to collecting data about completed tasks, productivity data could also be compiled through sensing technologies. Equipment and people could be tracked, and effective working time to complete tasks could be estimated. For example, sensors such as accelerometers and strain gauges could be installed on construction equipment and the activities inferred through machine learning (Soman et al., 2017; Harichandran et al., 2021).

What software is needed?

A central server is needed for collecting the information, performing analytics, and converting the information into useful and convenient forms that are amenable to management decision-making.

What are the technical challenges?

Communication links from remote construction sites may not be good, especially where there are many obstructions to wireless signals. A robust local network might need to be established on site.

What are the barriers?

Many monitoring and communication devices installed on site would increase the project cost. Ensuring proper functioning of these on site requires additional personnel.

How to quantify the benefits and effectiveness of automation?

Data about the time spent in manual inspections should be compiled. Also, the time taken for extracting required information from the reports should be estimated. With this, the time saved through automated progress monitoring can be calculated.

5.4 Summary

This chapter explores the possibility of applying automation techniques to on-site construction activities. An approach to analyzing the potential for automation is presented. The approach involves answering a series of questions to bring out current practices and benefits of automation. Twelve examples of construction activities are discussed to illustrate the application of the approach. It is shown that both hardware and software are equally important in improving the level of automation. Hardware design involves studying the kinematics and dynamics. High-level software for automation might require advanced computational techniques such as machine learning.

Even though it is feasible to apply automation and robotics to many of the examples that are discussed in this chapter, top level integration is essential for the best project performance. A networked system of automated machines might be needed, instead of many standalone single task robots (Bock and Linner, 2016a).

References

Bock, T. and Linner, T. (2016a). *Construction Robots: Elementary Technologies and Single-Task Construction Robots, Construction Robots*, New York, NY: Cambridge University Press.

Bock, T. and Linner, T. (2016b). *Site Automation Automated/Robotic On-Site Factories*, New York, NY: Cambridge University Press.

Harichandran, A., Raphael, B., and Mukherjee, A. (2021). A Hierarchical Machine Learning Framework for Identification of Automated Construction Operations, *Journal of Information Technology in Construction*, 26, pp. 591–623.

Krishnamoorthi, S. and Raphael, B. (2018). *A Methodology for Analysing Productivity in Automated Modular Construction*, ISARC 2018–35th International Symposium on Automation and Robotics in Construction and International AEC/FM Hackathon: The Future of Building Things.

Krishnamoorthi, S. and Raphael, B. (2021). *Performance Evaluation of Automated Construction Processes*, Built Environment Project and Asset Management. https://doi.org/10.1108/BEPAM-03-2021-0059

Soman, R.K., Raphael, B., and Varghese, K. (2017). A System Identification Methodology to Monitor Construction Activities Using Structural Responses, *Automation in Construction*, 75, pp. 79–90.

Part II

6 Introduction to machine learning

6.1 Introduction

Even though smart robots have been developed that perform several tasks better than humans, there are many activities that are still too difficult to automate. Working in confined spaces with limited access, actions that require hand-eye coordination to plan the precise movements to be made, etc. are challenges to machines. Humans have a huge amount of general background knowledge that helps them reason in new situations and find solutions that machines are still not capable of doing. However, recent progress in artificial intelligence provides indications that there will be new generations of machines having unimaginable capabilities. Machine learning is an area of computer science that automation engineers cannot ignore. The area was developed based on the realization that computers could learn and perform tasks better by mimicking human learning process. The term *machine learning* is commonly meant to denote the process of developing, testing, and implementing methodologies that help computers do tasks better using data (Raphael and Smith, 2013).

Previous chapters discussed the role of sensing and data analytics in implementing smart automated systems. When a system is being monitored continuously, large amount of data will be collected. Making sense of this data is not easy (Raphael and Harichandran, 2020). Most data sets contain noise and anomalies. These must be filtered out first. Then raw data should be converted into useful information for decision-making. For example, a machine-learning system might examine data collected from accelerometers installed on a construction equipment and determine that it could result in unsafe conditions; it could then take actions to prevent accidents (Harichandran et al., 2021).

Machine learning was developed as a specialized area of artificial intelligence (AI). AI explores processes that mimic human intelligence and includes broader areas such as logical reasoning, theorem proving, and computational neuroscience. Machine learning has more focus on learning from data. However, the boundaries of this area are blurred, and it is not always possible to clearly state whether a topic falls under machine learning or under the broader area of AI.

To understand what machine learning aims to achieve, it is useful to examine the process of human learning. Human learning takes place either by acquiring knowledge from sources such as human experts and books, or by self-discovery. We do not need anyone to tell us that falling from a height is painful if we have experienced it ourselves. Improvement of performance is an indicator of learning. Humans learn from experience and adapt their actions for future tasks. If a construction worker falls through a temporary uncovered opening in a floor slab, the site manager will immediately take steps to install guard rails at all similar openings to avoid repetition of the accident. Traditional computer programs do not exhibit this aspect of human behavior, that is, learning from experience. They produce the same output for the

DOI: 10.1201/9781003165620-8

same input all the time (unless they make use of random numbers for computing the output). They do not collect feedback about whether the output is acceptable to the user, and they have no mechanism for incorporating the feedback in the process of computing the output in the future.

Since the 1950s, researchers have been trying to develop techniques for making machines learn like humans. It was thought that we can develop models of human learning and use them for making machines mimic human capabilities. Later, the focus shifted to acquiring knowledge and learning from data, instead of mimicking human learning processes. In the 1970s many researchers experimented with the idea of representing knowledge in the form of rules that are elicited from experts and using them for decision-making. This has been only partially successful, and today there is not much interest in this technique. Then researchers realized that compiling and reusing cases is more promising than maintaining large knowledge bases containing precise and accurate rules. The Case-Based Reasoning (CBR) approach became popular as a result of this observation (Kumar and Raphael, 1997). Discovering knowledge from data looked promising when computers became powerful enough to manipulate large volumes of data. This approach has been immensely successful in certain applications such as recognition systems and natural language processing. Today, achievements such as accurate face recognition and speech-to-text translation do not surprise anyone.

Despite advances in machine learning, practical applications are rare in the civil engineering domain. We have not fully exploited the potential of many discoveries in computer science for improving our construction processes. This chapter introduces some of the machine learning algorithms and their possible applications in construction.

6.2 Machine learning tasks

Human beings learn from examples and experiences by compiling the lessons conveyed by them and checking whether they conform to the knowledge they already have. They modify their knowledge in the light of the new information and use that in future decision-making. Machine learning follows a similar approach. Data are provided to the system in the form of training examples. Knowledge is extracted from the training data and generalized such that it can be reused in situations that are not exactly the same. However, machine learning requires a precise definition of the learning task. There are mainly three types of learning tasks:

a Supervised learning
b Unsupervised learning
c Reinforcement learning

In supervised learning, each data point in the training data consists of values of input and output variables. You need to define what are the input and output variables a-priori. The task is to find the relationship between input and output variables. The relationship is learnt by searching among a predefined set of possible relationships, known as the *hypothesis space*. Learning typically takes place by adjusting the values of parameters of a model that computes the output from the input. For example, a simple relationship is a linear equation in which the coefficients of variables are unknown. In practical applications, relationships are non-linear and more complex. After learning this relationship, it is used for predicting the values of output variables when only the values of input variables are known.

In unsupervised learning, variables are not separated into input and output; all the variables are treated equally. The training data contains the values of all the variables. The learning task is to identify trends and patterns in the data, which could be used for making inferences. For example, the knowledge that the data follows a Gaussian distribution with a certain mean and standard deviation is useful for detecting anomalies. If a new data point is obtained that does not follow this trend, it is inferred that the point displays anomalous behavior. Another commonly used unsupervised learning method is clustering. Clustering group data points that are close to each other helps identify patterns in the data set. It is a way of synthesizing knowledge into a form that is easy to understand and reuse. For example, if the construction productivity data for different contractors in a project is found to fall into two clearly defined clusters, it could mean that the contractors in the two clusters might be following very different systems or processes or using different types of equipment. You may not be very successful if you look for a common relationship between variables among all the data points. It will be more effective to consider data points in each cluster separately for discovering good relationships between variables.

Reinforcement learning aims to identify the optimal sequence of actions to achieve a goal in a complex environment. The simulation environment is defined by specifying possible actions that could be taken, that is, by defining the rules of the game. Feedback is provided to the learning algorithm in the form of penalties and rewards. A reward function is defined by the user that indicates how good is the current state compared to the goal state. The assumption is that even though we are not able to clearly say how good an action is in a given situation, after an action is taken, we will be able to evaluate the state and say whether an improvement has been achieved.

Consider a 3D-printing problem. The printer must deposit material into all the grayed cells shown in Figure 6.1, in a single pass. That is, the printer head should move continuously from one cell to the next without stopping or jumping to a cell that is far away. At each cell, the printer head can move in the x or y direction to the neighboring cell. These are the rules of the game. The output we want from the machine-learning module is the sequence of steps to be followed. The task is completed when all the cells are filled. There are many possible paths to be explored by trial and error. The reward function is defined such that the highest value is returned when all the cells are filled, and highest penalty is given whenever there is a deadlock. The reinforcement learning algorithm identifies the best sequence of actions by trial and error using feedback through the rewards function.

Reinforcement learning is an advanced topic in machine learning and is a subject of intense research. The application of reinforcement learning to practical engineering problems involves several challenges, such as precise representation of all the rules and operators. Hence it is not discussed further in this book which is meant mainly to introduce machine learning applications for construction and building automation.

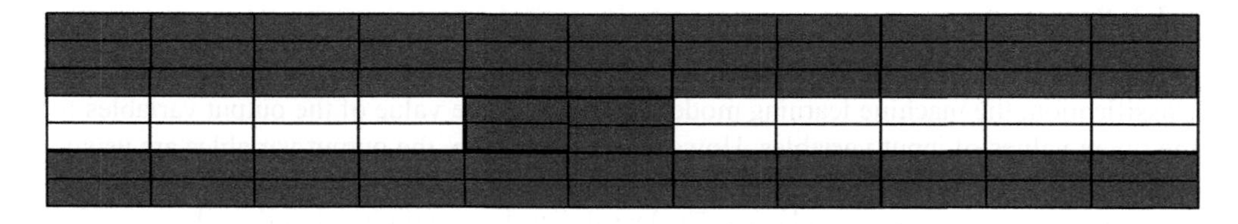

Figure 6.1 Path planning for 3D printing

Creating a machine learning model requires a precise definition of the learning task. This involves specification of the input and output variables as well as the hypothesis space. In addition, several hyperparameters of the model must be tuned to get good results. We are still far away from an ideal situation in which all the learning takes place automatically without much interaction with the user.

6.3 Supervised learning

Two important supervised learning tasks are classification and regression. This section provides an overview of these tasks; subsequent chapters discuss these topics in more detail.

6.3.1 Classification

In classification tasks, the output variables are class labels, which take discrete values. The value of the output variable indicates to which group (class) the data point belongs. For example, a worker on a construction site might be idle or performing heavy work. These states could be represented by the labels "idle" and "working". Imagine that a construction company has put sensors on workers, and the values read by these sensors are available in the form of a time series. The company wants to find out whether the worker is idle or working, using only the sensor readings. The training data can be visualized as a two-dimensional table in which each row contains data at a specific time. Each row is considered as a data point that consists of the values of input variables (sensor reading) along with the output variable (class label). In supervised learning, the class label (output variable) has to be supplied along with the input variables. The machine-learning algorithm determines which features in the input data separates the classes "idle" and "working". Once this relationship is learnt, it can "predict" whether the worker is idle or working by looking at the sensor data. Classification might seem like clustering, but it is a different type of learning task. Clustering is an unsupervised learning tasks in which the clusters are determined automatically, whereas, in classification the class labels are provided by the user in the training data set.

A popular application of classification is image recognition. Suppose a mobile robot on a construction site wants to locate a particular piece of equipment. A machine-learning module could be implemented for achieving this. Several images of the equipment are presented to the module with the label "equipment"; these are considered as positive examples. To help the module learn features that are not present in other images, a set of negative examples are also presented. The algorithm learns what are the common features present in the positive examples that are missing in the negative examples. Once these features are learnt, it can tell whether the equipment is present in a new image. That is, it is able to predict the class label by examining the data.

6.3.2 Regression

In regression, the objective is to learn a relationship between input and output variables. Like classification, the machine learning model must predict the value of the output variables for the given values of input variables. However, in regression, the output variables are usually continuous, not discrete. Therefore, it does not make sense to discuss what separates the data points into different groups or classes. The learning task essentially minimizes the difference between what is predicted by the model and the corresponding value of the output variable given in the training data. This is an optimization problem in which the decision variables are the model parameters, and the objective function is the prediction error. For example, consider the task of controlling the flow of treated air into a room for maintaining thermal comfort. Assume that the air temperature in the room is 27°C. How much air at 18°C should be supplied to the room so that the room temperature reaches 24°C? If all the environmental parameters are known and an accurate model is created, a detailed thermal simulation will provide the answer. Suppose such a model is not available. How will you compute the volume of air needed to achieve the target temperature? This can be done by training a statistical model using past data. Here, the output variable is the volume of air, and the input variables are the temperatures and other known environmental parameters. Through regression, a relationship (possibly a non-linear equation) is found between the input and output variables by adjusting the parameters in the model such that the error between predicted and actual values is the minimum. It should be noted that the mathematical model need not be restricted to simple polynomials that you are familiar with. Artificial neural networks (ANN) are more complex mathematical models that have been successfully used for regression tasks. The mathematical model in an ANN is a network of simple relationships that are combined using weight factors. However, the fundamental principle is the same – minimization of error by selecting the best values of model parameters. There is an alternative viewpoint to error minimization, based on Bayesian probability; this is discussed in detail in the book by Bishop (2006).

The use of complex models such as ANN helps to accommodate a mixture of discrete and continuous variables for accurate prediction of the output variable. Regression trees and random forests are other algorithms that are good at this. Sometimes the form of the equation changes completely based on the value of a discrete variable. For example, the parameters that affect the productivity of construction equipment might be different based on the type of the equipment and the source of energy (diesel or electricity). In general, we may not know a-priori which variables influence the form of the equation. It must be inferred automatically from data. Such tasks are easily handled by random forest algorithms.

6.3.3 Generalizability

How do you make sure that the relationship that has been learnt is correct? What if there is a spurious relationship in the training data (by accident or because of bias in sampling)? This issue was made popular by a statistician who presented an interesting graph linking divorce rate in the US with the per-capita consumption of margarine. Using US Census data, he showed that the divorce rate is related to the consumption of margarine with a correlation coefficient of 0.99! The message from this study is that correlation is not equivalent to causation. That is, a high correlation between variables need not necessarily mean that one causes the other. Such spurious relationships cause difficulty in learning correctly from data. Since machine-learning algorithms do not have complete domain knowledge and lack the ability to reason about the relationships, it could discover wrong relationships that connect input and output variables. This interesting example is often quoted in lectures on image recognition: In an experiment involving identification of images of cars, all the images of cars contained trees in the background by pure coincidence. The machine-learning algorithm identified the common pattern in all the images of cars, which happened to be trees, not cars. When a picture of a tree was shown, the algorithm wrongly inferred that there is a car in the picture. It might be argued that while selecting the data for training, we should make sure there are no images of other objects. However, ensuring there are no spurious patterns in data are difficult in practice because human beings are not capable of recognizing all the patterns in the data. A machine might identify patterns that humans cannot see. For example, if a single pixel is common in all the images of a particular class and if this pixel value is different in all the remaining classes, a learning algorithm might identify this pixel value as a discriminating feature that separates the classes. In the case of regression, a particular combination of parameters might result in the function passing through all the data points with zero error. But it might not predict correctly in new cases. We call this effect *overfitting*. See Figure 6.2. Values of two variables p1 and p2 for four data points are plotted along x and y axes. The thick dotted line represents the average line passing through all the points. This is the equation you would get if you performed linear regression. The curved shape shown in the figure is a second-degree polynomial obtained as the trend-line using a spreadsheet program. The spreadsheet program tried to minimize the error in every data point and calculated the coefficients of the polynomial. This resulted in a curve that has large deviations from the average line. Even though, the errors in the training data (the four data points shown in the figure) are small, the errors are likely to be large in the case of "unseen" data points, that is, the points that are not used in the training. This is an example of overfitting.

The problem is compounded by the noise in the data. Real data always contain noise. Measurements are never accurate or precise. Observations are influenced by human errors or mistakes in interpretation. In general, statistical variations are possible because of factors that are beyond our control while collecting data. The presence of such noise in the data might cause a learning algorithm to fit a wrong relationship to the observed data such that noise strongly determines the form of the relationship.

If large volumes of data are available, the chance of learning spurious relationships is reduced. The probability that an irrelevant feature is present in all the data points becomes small as the number of data points is increased. This raises the questions: How many data points are needed? How are we sure that the system has learnt correctly? How do we make sure that the system predicts correctly in the case of unseen data, that is, data that was not used for training and is probably very different from the training data?

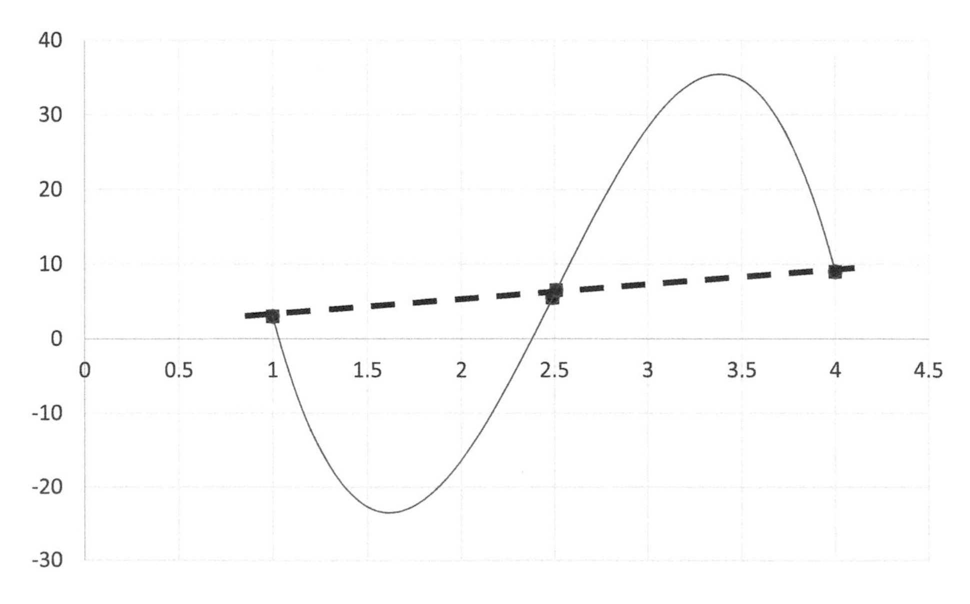

Figure 6.2 Fitting a straight line and a second-degree polynomial to the same data. The x and y axes represent two arbitrary variables p1 and p2.

The ability to predict correctly in the case of unseen data are known as generalizability. This requires that the learning algorithm should use generic features that are likely to be present in the population representing the data sample. It should not use specific patterns that are present only in the training data. But how do we find out what patterns are generic and what are likely to be specific features of the data set? The concept of Occam's razor is useful here. Assume that there are N data points, and an algorithm learns N rules to identify each data point. If a new test data point matches a training data point exactly, the corresponding rule will be able to classify the point correctly. However, if the data point is slightly different, all the rules might fail. We say that such a system lacks generalizability and tends to overfit. This is because a system containing N rules obtained from N data points is unnecessarily complex. To avoid overfitting, it is necessary to select the simplest possible relationship to explain the data. This principle is called Occam's razor. For example, a straight line might be better than a tenth-degree polynomial that fits all the data points exactly. The high-order polynomial might be influenced by the noise in the data and may not predict accurately in the case of unseen data.

6.3.4 Training, validation, and testing

The terms *validation* and *testing* are often used interchangeability. However, it is useful to make a distinction between the two. In supervised learning, the term *test data* is used to denote a set of data points that are used to assess the performance of a model that has been fully trained. Usually test data are different from training data. The term *validation data* is used to denote a set of data points that have not been used for training and are used to evaluate the performance of a model during the process of training. The validation data might be used to tune the parameters of the model. For example, the performance of a linear model might be evaluated using the validation data, and a decision is taken whether to go for non-linear models in the next stage of training.

Cross-validation is a technique that is used for checking generalizability, particularly when multiple models need to be tested for the best prediction capability. Many algorithms involve options to select from among different types of models, for example, the degree of polynomial to be used for regression. In addition, the models contain hyperparameters whose values are fixed by trial and error, and training must be done many times using different combinations of parameters. You need some *unseen* data to test whether the algorithm has learnt correctly using a set of hyperparameters.

In the cross-validation technique, a certain percentage of training data are kept aside for validation. This data are called validation data and is not provided as input data during the training phase. The system is trained using the remaining points; that is, validation data are kept aside as unseen data. After each phase of training, validation data are used to check the prediction accuracy, and hyperparameters are adjusted if necessary. Training is continued until the required accuracy is obtained. After the training is completed, a third data set, called the test data, is used to evaluate the performance of the system. In practical applications, the test data are available only during the actual use of the system. However, for research purposes, it is common to keep aside a part of the data for testing so that performance of different algorithms can be benchmarked.

In *S-fold cross validation*, the available data are randomly grouped into S "folds" or groups. Only (S − 1) groups are used for training. The last one is used for validation. Training runs are done for different combination of folds. A special case of S-fold cross validation is *leave one out* scheme in which the number of folds (S) is equal to the number of data points (N). In that case, only one data point is kept aside in a run for validation.

A difficulty with cross-validation is that it is expensive. When data are scarce, if a few points are left out for validation, enough data may not be available for training. Therefore, theoretical analysis is important for checking whether the model has sufficient generalizability and whether the number of data points is sufficient. Using a machine-learning toolkit as a black box is dangerous! Deeper understanding of the algorithms is necessary for selecting the right model and sets of model parameters.

One measure of the performance is the percentage of points in the test data in which the system predicts correctly, that is, the prediction accuracy. However, it does not give detailed information about where there are more mistakes in prediction. A confusion matrix helps to provide such information. An example is given in Table 6.1.

In this example, 100 data points are used for validation in a binary classification problem, out of which 80 belong to the negative class and 20 belong to the positive class. Each row in Table 6.1 represents the actual numbers belonging to each class. Each column corresponds to the number of predicted values belonging to each class. In the first row, the first column contains the number of data points that are predicted correctly by the model as negative, 40. The second column contains the number of data points that are wrongly predicted as positive. The sum of these two columns in the first row is the total number of points that are actually negative.

Exercise 6.1

- a What is the overall accuracy of the machine learning model according to Table 6.1?
- b What is the percentage of false positives? (points that are wrongly classified as positive)
- c What is the percentage of false negatives? (points that are wrongly classified as negative)?
- d What is the advantage of using the confusion matrix in addition to overall accuracy?
- e Which is better when the overall accuracy is the same: more false negatives or false positives?

Answer

- a The diagonal elements in Table 6.1 represent correct predictions. Forty are correctly predicted as negative, and 15 are correctly predicted as positive. Therefore, overall accuracy is 55%.
- b Out of 55 points that are classified as positive, 40 are false. That is, 72.7% are false positive.
- c Out of 45 points that are classified as negative, 40 are correctly classified. Five are wrongly classified as negative. Therefore, the percentage of false negatives is 11.11%.
- d For the same overall accuracy, the percentages of false positive and negatives could be different. This information is not available if only the overall accuracy is reported. A confusion matrix helps to visually evaluate which classes have more prediction error.
- e Whether more false positives are better than false negatives depend on the application. In some cases, false positives require expensive operations to verify the results, and the

Table 6.1 Illustration of a Confusion Matrix

(n = 100)	*Predicted: Negative*	*Predicted: Positive*
Actual: Negative	40	40
Actual: Positive	5	15

cost may not be acceptable. In other cases, false negatives might result in dangerous conditions. For example, if a piece of equipment is wrongly classified as being in good working order, it might cause accidents.

6.4 Unsupervised learning

Popular unsupervised learning tasks in engineering include clustering, identifying relationships and factors among variables, and density estimation. Here, distinction is made between learning tasks and algorithms. Tasks indicate what we want to achieve, whereas algorithms denote how these tasks are accomplished. Unsupervised learning algorithms are described in Chapter 11.

6.4.1 Clustering

The goal of clustering is to group together data points that are similar. This is an unsupervised learning algorithm. That is, the class labels are not provided for training data, unlike in the classification task. The algorithm automatically assigns each data point to a cluster based on its distance to the cluster. After the clusters are formed, experts might interpret the results, examine the characteristics of clusters, and give each cluster a label based on their interpretation of the cluster. For example, an expert might recognize that all the points in one cluster represent normal behavior, and another cluster contains anomalies. When a new test point is obtained, its cluster is predicted based on its distance to the clusters. Using common characteristics of the cluster, the behavior of the new point might be predicted (Saitta et al., 2008).

Clustering has applications in pattern recognition, image processing, and classification. Trends and patterns in data are found by clustering. It is most useful when data consists of many attributes since humans are not good at seeing patterns in multiple dimensions.

Clustering helps to simplify the presentation of data such that users can get a concise summary without too many details. For example, search engines group together documents that are similar and form clusters. Small variations within a cluster may not matter much, and the presentation will be simpler.

A central concept in clustering is similarity, which is when all the points within a cluster are supposed to be similar. Similarity is determined using a measure of distance known as the metric. A simple metric is the Euclidean distance, which is the n-dimensional equivalent of the geometric distance that we are familiar with. The distance between two points in three dimensions is the square root of the sum of squares of the distances along the three directions. In n-dimensions, there are more variables; the procedure is simply repeated for all the variables and is written as follows:

$$d(\mathbf{x}_1, \mathbf{x}_2) = \left(\sum_{i=1}^{n} \left(x_{1,i} - x_{2,i} \right)^2 \right)^{\frac{1}{2}} \qquad \textit{Eq 6.1}$$

Here, d is the distance between the two data points x_1 and x_2; $x_{1,i}$ and $x_{2,i}$ are the values of the i-th variable in the two data points; n is the number of variables. This formula is commonly known as the second norm of the vector. This formula gives equal weight to all the variables. If the scales and units are different, it is important to normalize the data before any analysis is performed. A standard way of normalizing data are to divide each value by the standard deviation after subtracting from the mean. The following formula does that:

$$x'_i = \frac{(x_i - \overline{x}_i)}{\sigma_i} \quad \text{Eq 6.2}$$

Here, x'_i is the normalized form of variable x_i; $\overline{x}_i$ and σ_i are the mean and standard deviation of this variable. This formula converts the variables into a new unit, which measures how many standard deviations away from the mean is each value. There are other ways to normalize data, for instance, by dividing by the maximum range of values after subtracting from the minimum value. This ensures that all the values are between 0 and 1. If some variables are more important in assessing similarity, it is necessary to take a weighted sum of the distances as follows:

$$d(\mathbf{x}_1, \mathbf{x}_2) = \left(\sum_{i=1}^{n} w_i \left(x_{1,i} - x_{2,i} \right)^2 \right)^{\frac{1}{2}} \quad \text{Eq 6.3}$$

Here, w_i is the weight for the i-th variable. Once a distance metric is selected, points can be grouped together such that those that are close by are in the same cluster (group). Usually, the number of clusters is predefined and is given as an input. Then, the task is to assign each data point to one of the clusters in order to maximize some score that evaluates how compact the clusters are. In general, the average distance between the points within the same cluster should be minimized, and the distance between the clusters should be maximized. This helps to identify clusters that intuitively appears distinct on visual inspection. However, note that visually examining clusters is possible only in two or three dimensions. In higher dimensions, techniques to reduce the number of relevant variables must be considered. Principal Component Analysis is a technique for dimensionality reduction (Section 6.4.3.2).

Exercise 6.2 Group discussion

In Figure 6.3, the amount of excavation completed by different contractors for a hypothetical airport construction project is plotted as a function of the day. Discuss the following:

a Are there clusters in this data?
b How many clusters are there? (assuming the answer to the first question is yes)
c What are the characteristic features of each cluster?
d What is the reason for the presence of clusters from a project management viewpoint?

Since the units for x and y axes are different, it is important to normalize the variables before clustering is performed. The mean and standard deviation of each variable are computed, and the values are normalized as explained previously. The normalized variables are plotted in Figure 6.4. It can be seen that the patterns are not affected after normalization. However, the points that lie above and below the average are clearly visible. A group of points are seen lying below the $y = 0$ line (mean value), well separated from the remaining points on the same day. It seems that there are two groups of companies, one with high productivity rates[1] and the second with low productivity rates. They might be using different types of equipment and processes for the same work. Probably, they were just unfortunate to have selected bad terrain where the work is difficult.

Weather was bad on some days and a general dip in productivity rates is noticeable for those days. Even on those days, there are two distinct levels of productivity. The apparent

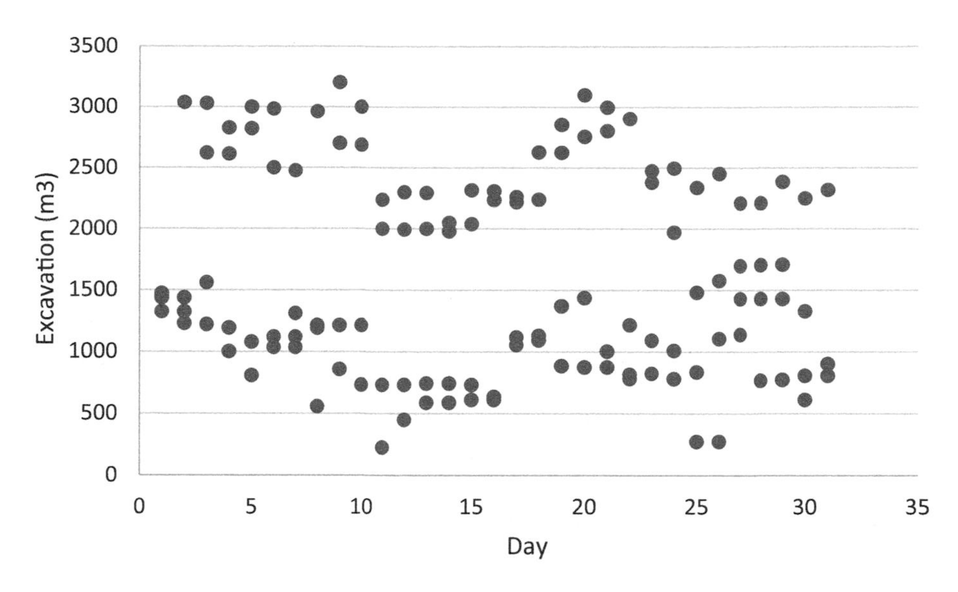

Figure 6.3 Volume of excavation completed per day by different contractors

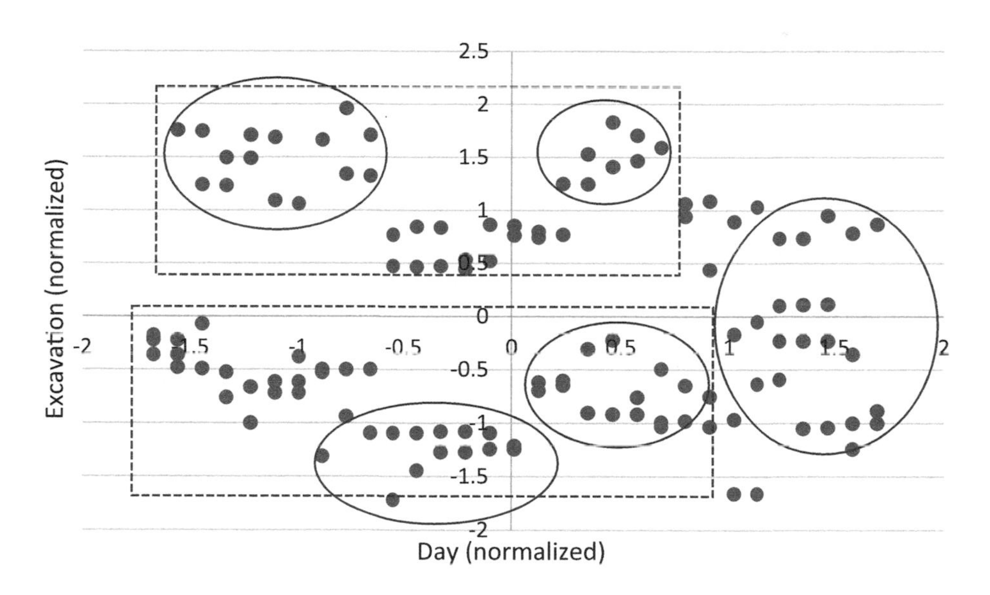

Figure 6.4 Normalized form of the data shown in Figure 6.3. Oval indicates clusters. Dotted boxes show clusters that are obtained by merging smaller clusters

clusters in the data are a result of the combination of bad days and low productivity of some companies. Possible clusters are shown as circles in Figure 6.4.

After identifying clusters in the data, it will be useful to examine other variables, which were not used for clustering, to see how they affect the formation of the clusters. It is not advisable to include all the variables in any machine-learning process since it will adversely affect the efficiency of the process and the accuracy of the results. Engineering judgement is usually needed to select the right set of variables for obtaining the best results.

6.4.2 Density estimation

The word density in "density estimation" refers to probability density. We want to find out what is the probability of getting a particular value for a variable. In the case of continuous variables, we calculate the probability density, which is the probability per unit value of the variable. If large amount of data are collected during the actual operation of engineering systems, probability distributions could be estimated using Bayesian updating (see Chapter 7).

There are several uses for the probability density data. When we collect data, if we find values with low probability, we need to suspect the validity of data (Harichandran et al., 2021). It could indicate an anomaly or outlier. We can filter out noise in data by computing probabilities. Another use is that we can theoretically compute the probabilities of other dependent variables using the probabilities of independent variables (Domer et al., 2006; Raphael et al., 2007).

Probability densities and other statistical parameters are useful characteristics of data that could be used for decision-making. In this form of unsupervised learning, learning involves generating additional information from data that represents generalized knowledge related to the domain or the application. Thus, it is a process of knowledge discovery.

6.4.3 Identifying relationships among variables

Two techniques are discussed in this section: a) correlation and b) principal component analysis (PCA).

6.4.3.1 Correlation

Correlation indicates which variables are related. Sometimes we observe that values of certain variables increase or decrease together. That is, if the value of one variable is high in a data point, the corresponding value of the second variable is also high. A reason for this might be that one causes the other (causality). Another possibility is that there is an underlying factor (which cannot be directly observed or measured) that determines the values of both the variables. Whatever might be the case, it is helpful to determine whether such relationships exist in the data. It is a form of useful knowledge. The equation to compute correlation is given in Section 7.3.

EXAMPLE

Different contractors were assigned different parts of the site for an excavation task. Some encountered hard ground on some days, and some found soft ground; these factors might have affected the productivity. The data that was collected from a project is shown in Table 6.2. The project manager is interested in finding out which variable is most strongly related to productivity. This information is obtained by computing the correlation between productivity and the other variables.

Correlation between variables is an important characteristic of the data set. The probability distribution of dependent variables is affected by such relationships. If simulations are performed using synthetic data, we need to ensure that correlation between variables is preserved in the data set that is used.

Table 6.2 Productivity Data in an Excavation Project

Day	*Contractor*	*Productivity*	*Shift*	*Ground*
1	1	700	day	hard
1	1	660	night	hard
1	2	740	day	soft
1	2	720	night	soft
2	1	730	day	hard
2	1	670	night	hard
2	2	720	day	soft
2	2	680	night	soft

In addition to providing useful knowledge about the data, correlation can also be used in interesting ways for practical tasks. For example, it has been used in fault diagnosis. It is observed that certain variables are strongly correlated under normal functioning of a system. However, when faults develop, this relationship is changed. A change in the correlation coefficient might be a result of a change in the system state and could be used to detect faults.

It should be remembered that correlation can only measure the linear relationship between variables. Even if there is a strong relationship, if it is non-linear, the correlation coefficient might be low. For example, imagine if there a cosine relationship between the two variables x1 and x2 as follows:

$$x2 = \cos(x1)$$

If values of x1 are randomly generated between 0 and 2π, and the corresponding values of x2 are calculated using the previous equation, the resulting correlation between the two variables will be close to zero. Therefore, the absence of correlation need not indicate a null relationship between the variables. It could be because of a non-linear relationship between the variables.

Spearman's rank correlation might be used to determine non-linear relationships that are monotonic, that is, when both variables increase or decrease together without having a linear relationship between them. In Spearman's rank correlation formula, the rank of each value in each data point is used to calculate the correlation coefficient instead of the actual value of the variable. The rank of a value is its position in a sorted list of all the values. The rank correlation coefficient is better in detecting non-linear relationships because the ranks for values of variables increase or decrease together if the values increase or decrease monotonically.

6.4.3.2 Principal component analysis

The quality of surface finish on a construction job was evaluated and studied as a function of environmental parameters. The plot of surface finish with respect to air temperature is shown in Figure 6.5. What pattern is observed in the data? How can such patterns be learnt through unsupervised learning?

The surface-finish parameter in Figure 6.5 is roughly linearly related to air temperature, with some scatter around the mean line. In two dimensions, we are able to visualize the data easily and identify such patterns. Correlation gives an indication of such relationships for a pair of variables. However, when there are many variables, visualization becomes a challenge. Principal Component Analysis is a technique that could help in such cases. Principal components represent directions in which data has maximum variation in the given set. In

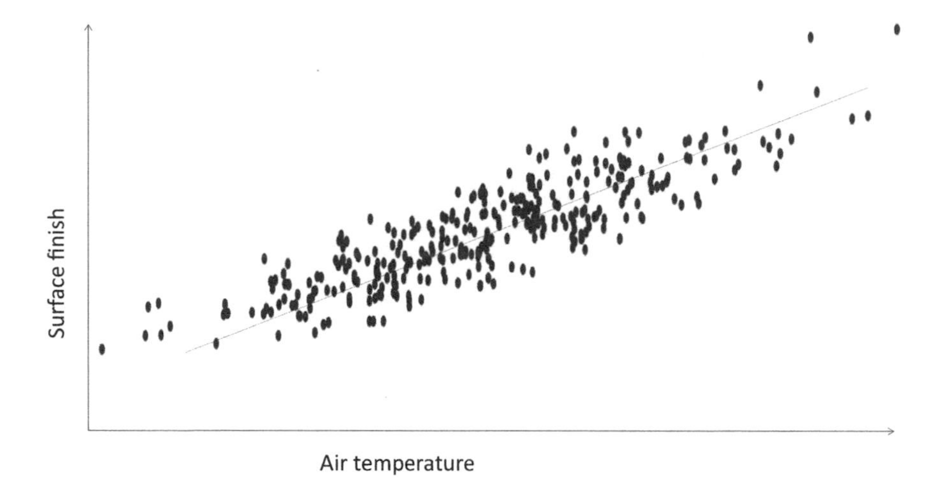

Figure 6.5 Scatter plot of correlated variables

Figure 6.5, the dotted line shows the direction in which the data has maximum variation. If you take a direction perpendicular to this line, you will notice that the data does not have much variation.

Principal components are eigenvectors of the covariance matrix of the data set (See Chapter 11). The corresponding variances in these directions are given by the eigenvalues. The eigenvectors are arranged in the decreasing order of the eigenvalues. The first eigenvector corresponds to maximum variance in data, and the last one has minimum variance. If the eigenvalues become too small below a threshold, those eigenvectors represent directions in which there is no significant variance in data. These directions can be omitted and ignored without any loss of information. Thus, PCA helps in dimensionality reduction. We are able to reduce the number of variables by ignoring directions in which the data does not have much variability.

6.5 Summary

Automation of engineering tasks involves collection, pre-processing, and analysis of large amount of data. Machine-learning techniques help in extracting knowledge from data and using it for intelligent control and operations. Currently machine learning techniques are mostly successful for certain types of tasks that are well defined. These tasks are introduced in this chapter. Details of algorithms for performing these tasks are covered in subsequent chapters.

Additional exercises

The objective of the following exercises is to develop skills to identify the nature of the machine-learning task and to formulate it as a well-defined problem. Attempt to define the problem as clearly as possible by specifying what are the variables, what data are needed for learning, what machine-learning techniques could be used for learning, etc. Some of the exercises require additional knowledge related to the specific engineering activity. Consult online material to collect as much information as possible so that the problem can be clearly defined.

1 Travel time prediction

An automated guided vehicle is used on a construction site for transporting material. We need to accurately predict the travel time of the vehicle from a given starting point to the destination. How can a machine learning problem be formulated to perform this task?

2 Construction cost prediction

Historic data related to construction costs for many projects are available in a database. For a new project, the total cost as well as the cost of individual components need to be estimated. How can the historic data be used for an accurate estimate?

3 Activity recognition

Accelerometers are placed on workers to collect real-time data. The patterns in accelerometer readings are used to predict the activity carried out by the worker. How can this be done using machine learning?

4 Vertical transport

A mobile crane lifts loads on a construction site using a system of hooks, pulleys, and cables. There is a limit for the load that can be safely lifted by the system. How can real-time data be used for the safe operation of the crane?

5 Alignment of formwork

Formwork made of timber is used for casting concrete walls for a building. If the formwork is not correctly aligned vertically and horizontally, the resulting wall will have defects and might have to be demolished and reconstructed. What sensors can be used to detect defects in alignment in real time? How can this data be used for ensuring that the work is carried out without any defects?

6 Rebound hammer test

Rebound hammer test is used to determine the compressive strength of concrete. The method involves impinging a mass on the concrete surface through a spring and measuring how much it rebounds. The rebound is related to the material properties. How do you develop and empirical model for predicting the strength of concrete from the rebound values?

7 Robotic drone

A robotic drone picks up material from a source and drops it at a specified destination. How can camera images be used to identify the precise location of the material to be picked up?

8 Bar-bending machine

An automated bar-bending machine takes as input a 3D model of the shape to be created. A straight bar is fed into the machine, and the machine bends it to the required shape. How can feedback be provided to the machine so that defects are corrected in real time?

Note

1 In project management literature, productivity refers to the amount of work done per time for unit input. Input is either manpower or other resources.

References

Bishop, C.M. (2006). *Pattern Recognition and Machine Learning*, New York, USA: Springer.

Domer, B., Raphael, B., and Saitta, S. (2006). KnowPrice2: Intelligent Cost Estimation for Construction Projects, *Lecture Notes in Computer Science*, 4200 (LNAI), pp. 147–152.

Harichandran, A., Raphael, B., and Mukherjee, A. (2021). A Hierarchical Machine Learning Framework for Identification of Automated Construction Operations, *Journal of Information Technology in Construction*, 26, pp. 591–623.

Kumar, B. and Raphael, B. (1997). CADREM: A Case-Based System for Conceptual Structural Design, *International Journal of Engineering with Computers, Springer-Verlag London Ltd.*, 13, pp. 153–164.

Raphael, B., Domer, B., Saitta, S., and Smith, I.F.C. (2007). Incremental Development of CBR Strategies for Computing Project Cost Probabilities, *Advanced Engineering Informatics*, 21, pp. 311–321.

Raphael, B. and Harichandran, A. (2020). Sensor Data Interpretation in Bridge Monitoring: A Case Study, *Frontiers in Built Environment*, 5, pp. 148. https://doi.org/10.3389/fbuil.2019.00148.

Raphael, B. and Smith, I.F.C. (2013). *Engineering Informatics: Fundamentals of Computer Aided Engineering*, second edition, Chichester, UK: John Wiley.

Saitta, S., Kripakaran, P., Raphael, B., and Smith, I.F.C. (2008). Improving System Identification Using Clustering, *Journal of Computing in Civil Engineering*, 22 (5), pp. 292–302.

7 Basic mathematics for machine learning

7.1 Introduction

Understanding machine-learning algorithms requires knowledge of basic concepts related to probability, statistics, and linear algebra. These are introduced in this chapter. An attempt is made to explain this concept using examples from building and construction domains. Simple numerical exercises are also provided for better understanding of the theory.

7.1.1 Inverse problems in engineering

Machine learning is used to solve difficult mathematical problems which cannot easily be solved using simple algorithms. Inverse problems belong to this class. The concept of inverse problems is introduced in this section.

Most engineering analyses aim to infer the effects (output) from the cause (input). Sometimes we have a direct mathematical expression that computes the output from the input. In other cases, we might have to solve differential equations. Since differential equations do not have closed form solutions in most cases, we must use numerical methods such as finite element analysis. In structural analysis, we calculate the internal forces and deflections from the loads. In simple cases, we can use algebraic expressions. In more complex cases and non-standard boundary conditions, we must use the finite element method. In thermal simulations, we compute the temperatures inside the building from the position of the sun and climatic conditions. Again, we must use computer software that employs numerical methods. In all these cases, we get unique answers that are accurate enough within the limits imposed by the uncertainties in input parameters and errors in numerical methods as well as the model of the physical phenomenon. That is, for one set of input we get one set of output.

However, if we wish to go back from the effects to the cause, it is not so straightforward. There could be multiple causes that produce the same effects. Therefore, the solution might not be unique. Computing the input from the output is known as the *inverse problem* and is more difficult than the forward problem. Solving a differential equation to compute the output from the input and boundary conditions is straightforward; However, it is not easy to infer the input conditions for the given output values (Vernay et al., 2015). In terms of fundamental logic, the process is called abduction (Raphael and Smith, 2013). Abduction is inherently unreliable, except in trivial cases where the closed-world assumption is valid. Nevertheless, many engineering tasks involve abduction: In design, we decide on the dimensions and properties that satisfy the constraints on output variables such as stresses

DOI: 10.1201/9781003165620-9

and deflections. In diagnosis, we attempt to find causes for observed failures. While solving complex problems, we know the end state, and we want to find the path to reach there. Expert engineers are paid high for their skill in performing abductive tasks. Now, we expect smart software systems to perform the same tasks.

Inverse problems are frequently encountered in automation and control tasks. A classic example is a three-degree-of-freedom articulated robotic arm. The tip of the arm has a tool (end effector) to perform tasks such as picking up objects or welding. The arm must move in the x, y, and z directions to perform the task. Internally, these movements are achieved by operating motors, which can produce only rotary motion. Given the current rotations (angles) of the joints, we can use "forward kinematics" to calculate the x, y, z positions of the end effector. However, if you want to find out what rotations will produce the required movements in the cartesian coordinate system, we must use "inverse kinematics". This is a more difficult problem due to many factors, including non-linearity of the inverse trigonometric functions, discontinuities and singularities, and the presence of multiple solutions.

Many control tasks involve solving inverse problems. Choosing the optimal control action requires an accurate assessment of the state of the system. In simple cases the parameters that characterize the state of the system can be measured directly. However, in many practical scenarios, it is either expensive or infeasible to install sensors for direct measurements. Instead, indirect measurements are relied upon. For example, imagine a robotic crane that is programmed to lift an object. The position of the lifting hook of a crane is needed to determine how much the arm has to be moved to lift the object. It may not be easy to install sensors on the tip of the arm to measure the distance. It will interfere with the operations of the arm. Instead, the position of the arm is inferred from the rotations of joints and other measurements. In some cases, this may be computed directly. But there are situations where the non-linearity of the system, environmental disturbances, and other uncertainties that cannot be quantified precisely, cause the computations to be inaccurate. In such situations, it is necessary to infer the state of the system by solving the inverse problem after measuring the system responses (Vernay et al., 2014; Papadopoulou et al., 2016). This task is known as *system identification*, an area where machine learning and statistical techniques are extensively used.

Several machine learning methods have been applied to inverse problems because they usually have high computational complexity (see Raphael and Smith, 2013 for a discussion on complexity). In fact, supervised learning tasks are essentially inverse problems. Learning algorithms attempt to infer models and model parameters from data that measure their predictions; that is, they try to infer cause from effects.

7.2 Elementary statistics

Many machine-learning algorithms have strong foundations in probability and statistics. Hence, these topics are discussed briefly here before going into the details of algorithms. Please refer to mathematics textbooks for a more detailed treatment of these topics.

Most engineering tasks involve manipulating variables whose values are not known precisely. When the values are not precisely defined, we treat them as random variables. Statistics deal with the analysis of random variables. This involves the application of probability theory. Even when there is uncertainty in the values of variables, we notice that certain values are more likely than others. We represent this information using a probability distribution.

There is a well-defined theory related to the computation of probabilities. Section 7.4 gives an overview of key concepts related to probability.

Common statistical parameters that are used in the analysis of data include mean, variance, and standard deviation, whose definitions can be found in elementary textbooks on statistics. Other concepts that are essential for understanding machine-learning techniques are described in the following sections.

7.2.1 Covariance matrix

The variance of a random variable quantifies the average difference of individual values in a sample from the mean. If all the values are equal to the mean (that is, there is no variation), the variance is zero. If the values fluctuate a lot, the variance is high. Covariance is a related concept; it measures the relationship between two random variables, that is, how much they vary together. The formula for covariance is given by:

$$Cov(x,y) = \frac{\sum_{i=1}^{N}(x_i - \bar{x})(y_i - \bar{y})}{N-1} \qquad \text{Eq 7.1}$$

Here,

$\bar{x}$ is the mean of the variable x, $\bar{y}$ is the mean of the variable y,
x_i and y_i are the values of x and y in the i-th data point, and
N is the number of data points.

The quantity $(x_i - \bar{x})$ measures the deviation of the i-th data point from the mean. The product $(x_i - \bar{x})(y_i - \bar{y})$ is positive if both the values of x_i and y_i are higher than their mean values. This product is positive also when both the values of x_i and y_i are less than their mean values. Therefore, the sum of this product over all the data points will be large if the values of x and y increase or decrease together. That is, high value of covariance indicate that the values of x and y are increasing or decreasing together; that is, there is a relationship between the two variables. If the value of y decreases every time the value of x increases, the covariance will be a large negative number.

It is useful to show the relationships between all the variables in the form of a matrix. This matrix is called the covariance matrix. The element at the j-th row and k-th column in this matrix is the covariance between the j-th variable and the k-th variable. It can easily be verified from Eq 7.1 that the diagonal elements of this matrix are the variance of each variable. That is, the covariance of a variable with itself is the variance of the variable.

Exercise 7.1

A contractor performs tasks related to brickwork and concreting at different locations within a site. Hourly productivity data are collected from the construction site. Productivity is expressed in non-dimensional units by dividing by benchmark values and tabulated in Table 7.1.

Table 7.1 Productivity Data

Hour	*Productivity rate – brickwork*	*Productivity rate – concreting*
1	1.045	0.545
2	0.982	0.464
3	0.928	0.536
4	0.964	0.428
5	0.964	0.518
6	0.964	0.518
7	1.153	0.491

Calculate the covariance matrix for the productivity rates.

Answer

The mean productivity rate for brickwork (x1) is 1.0 and that for concreting (x2) is 0.5. Subtracting the means from the two columns, the centered data are given as follows:

x1 – mean(x1)	*x2 – mean(x2)*
0.045	0.045
−0.018	−0.036
−0.072	0.036
−0.036	−0.072
−0.036	0.018
−0.036	0.018
0.153	−0.009

Multiplying by the transpose of this matrix, the covariance matrix is obtained as:

0.035	0.000
0.000	0.011

The variance of x1 is 0.035 and that of x2 is 0.011. The covariance between x1 and x2 is 0 (non-diagonal terms).

7.2.2 Correlation

Correlation was introduced in Chapter 6. Only the mathematical details are presented here. Correlation between two random variables x and y is usually computed using Pearson's correlation coefficient, which is defined as:

$$Cor(x,y) = \frac{Cov(x,y)}{\sqrt{var(x).var(y)}} \quad \textit{Eq 7.2}$$

Here,

$Cov(x, y)$ is the covariance of x and y according to the Eq 7.1;
var(x), var(y) are the variances of the respective variables x and y.

Since the covariance of a variable with itself is the variance, the correlation of a variable with itself is 1.0 (verify using Eq 7.2). The correlation between any pair of variables is bounded by −1 and +1 because the denominator contains the variances of these variables. A correlation of +1 indicates perfect linear relationship between the variables. A correlation of −1 also indicates a perfect linear relationship but with negative slope; that is, when one variable increases, the other variable decreases. High scatter about the mean line is indicated by low value of the correlation. A correlation coefficient of 0 means the variables are completely independent, and their variation is truly random with respect to each other.

Eq 7.2 can be written in a different form by substituting the expressions for covariance and variance as:

$$Cor\ (x,y) = \frac{\sum_{i=1}^{N}(x_i - \bar{x})(y_i - \bar{y})}{\left(\sum_{i=1}^{N}(x_i - \bar{x})^2 \sum_{i}^{N}(y_i - \bar{y})^2\right)^{1/2}} \qquad Eq\ 7.3$$

This form is useful in understanding why the correlation is bounded by −1 and +1. Consider two vectors **X** and **Y** that contain all the values of the variables x and y in the data set, after subtracting the mean values. Subtracting values from their mean is equivalent to centering the data by shifting the origin to the mean. The numerator in Eq 7.3 is the dot product of the vectors **X** and **Y**. The denominator contains the norm (length) of the vectors. Therefore, Eq 7.3 is equivalent to computing the dot product of two unit-vectors. If the two vectors are perfectly aligned, you get the maximum value of 1. You get −1 if the two vectors are exactly in the opposite direction.

Like the covariance matrix, we can compute the correlation matrix. This is useful to see which variables are strongly related. The correlation matrix is easier to interpret than the covariance matrix because the values lie within [−1,1], and the degree of interdependency between variables is more apparent.

7.2.3 Hypothesis testing

While looking for relationships in data, it is important to find out how strong is the evidence. We need to quantitatively establish how confident we are in stating that there is a certain type of relationship in the data. The concept of statistical hypothesis testing is often used for this.

Exercise 7.2

Productivity rates for the bar-bending activity was collected from a site. Manual tools for bar bending were initially used at this site. Then a mechanized bar-bending system was introduced. Data seems to indicate that there is higher productivity after mechanization. But how strong is the evidence? How confidently can we state that mechanization improves productivity?

The general procedure for hypothesis testing is to formulate two hypotheses. First, a null hypothesis states that a particular effect is not present. Then, an alternate hypothesis is formulated that the effect is present. We are interested in finding out the probability of our observations if the null hypothesis is valid. Sometimes we can calculate the probability of observations under the null hypothesis using logical reasoning. For example, if you toss an unbiased coin, you should get equal probability of heads or tails. In other cases, it may not be easy to compute probabilities using such reasoning, but data may be available, and this could be used for estimating probabilities. For example, in the bar-bending example described previously, the probability distribution of productivity before mechanization was introduced and can be computed from the measured data. To test a hypothesis, we compute the probability of the given observation (using actual data that is collected) under the null hypothesis. If the probability is too low, we reject the null hypothesis. Several test procedures have been developed for performing this scientifically. Popular techniques include t-test and z-test. (Readers are referred to statistics textbooks for details.) For illustration, consider one sample t-test. In this test, we calculate the difference between the actual mean of the data sample and the predicted mean under null hypothesis. Dividing this difference by the standard deviation gives a dimensionless number (which does not depend on the units of measurement). The probability of obtaining a specific deviation from the mean is a function of the sample size and can be computed using a standard function called the t-distribution. By choosing a threshold value for probability such as 0.05, we can accept or reject the null hypothesis. If the probability of getting the observed deviation according to the t-distribution is lower than 0.05, we can reject the null hypothesis with 95% confidence.

7.3 Linear algebra and data analytics

Examine the data shown in Figure 7.1. There are two variables x1 and x2. All the data points are strongly aligned in a particular direction shown as u1. The points are nearly collinear, with a few points showing small scatter around the average line. But this is not immediately evident if the raw data are tabulated with the values of x1 and x2, as shown in Table 7.2. When there are more than two variables, it becomes difficult to spot such relationships even with visualization, because collinearity is visible only if you select the right projections. How do you detect such relationships analytically? The correlation coefficient indicates whether two variables are related. But how do you determine collinearity in multiple dimensions? This section discusses this.

Exercise 7.3

1. Calculate the mean, variance, and standard deviation of x1 and x2, using the data given in Table 7.2.
2. Consider a new variable y1 = x1/4. What is the variance of y1? What is the relationship between the variance of x1 and y1?
3. When a variable is scaled by a factor *s*, how does the variance change?
4. Rotate x1 and x2 by 45° in the anticlockwise direction. Write down the expression for the coordinates (y1,y2) in the new coordinate system. Calculate the variance of y1 and y2.

Table 7.2 Collinear Data

x1	*x2*
2	0.51
3	0.75
5	1.25
−1	−0.25
−2	−0.5
−4	−1
−3	−0.76

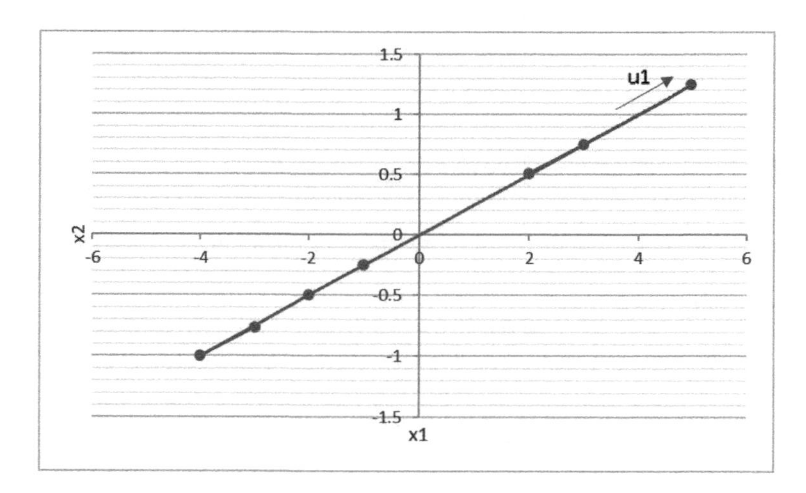

Figure 7.1 Data strongly aligned in one direction

Answer

1 x1: mean = 0, variance = 11.33, standard deviation = 3.366

x2: mean = 0, variance = 0.712, standard deviation = 0.842

2 The variance of y1 is 0.708. The variance of x1 is 16 times that of y1.
3 When the variable is scaled by s, its variance gets scaled by s^2.
4 The coordinates in the rotated coordinate system are calculated as follows:

From basic geometry, the unit vector inclined at an angle θ with the x1 axis has the coordinates (cos θ, sin θ). Therefore, the unit vector is written as:

$$\widehat{y1} = \begin{bmatrix} \cos\theta \\ \sin\theta \end{bmatrix}$$

Similarly, the unit vector in the direction y2 is given by:

$$\widehat{y2} = \begin{bmatrix} -\sin\theta \\ \cos\theta \end{bmatrix}$$

A data point P with coordinates (x1, x2) is written in vector form as:

$$\vec{P} = \begin{bmatrix} x1 \\ x2 \end{bmatrix}$$

The dot product of P and the unit vector gives the distance of the point in the direction defined by the unit vector. Therefore, the coordinate y1 in the rotated coordinate system is given by:

$$y1 = \vec{P}.\widehat{y1} = x1\ \cos\theta + x2\ \sin\theta$$

Similarly,

$$y2 = \vec{P}.\widehat{y2} = -x1\ \sin\theta + x2\ \cos\theta$$

These equations are written in the form of matrix multiplication as

$$\begin{bmatrix} y1 \\ y2 \end{bmatrix} = \begin{bmatrix} \cos\theta & \sin\theta \\ -\sin\theta & \cos\theta \end{bmatrix} \begin{bmatrix} x1 \\ x2 \end{bmatrix} = \boldsymbol{R} \begin{bmatrix} x1 \\ x2 \end{bmatrix} \qquad \text{Eq 7.4}$$

Where R is the rotation transformation matrix consisting of cos and sin terms as given in the previous equation. It is interesting to note that multiplying R with its transpose gives identity matrix I. That is, $R\ R^T = I$; In other words, R^T is the inverse of R. Using this expression, the rotated coordinates and their variances are given in Table 7.3 for $\theta = 45°$:

Table 7.3 Rotated Coordinates

	x1	*x2*	*y1*	*y2*
	2.000	0.510	1.775	−1.054
	3.000	0.750	2.652	−1.591
	5.000	1.250	4.419	−2.652
	−1.000	−0.250	−0.884	0.530
	−2.000	−0.500	−1.768	1.061
	−4.000	−1.000	−3.536	2.121
	−3.000	−0.760	−2.659	1.584
Mean	0.000	0.000	0.000	0.000
stdev	3.367	0.844	2.977	1.784
variance	11.333	0.713	8.865	3.181

The variance in the rotated coordinate system was computed numerically and presented in Table 7.3. Instead of numerically calculating the coordinates and computing the covariance matrix, an analytical expression for the covariance matrix can be derived using Eq 7.4. While doing this algebraic manipulation, a small change in the data representation should be noted. In our data matrix X, each row consists of one data point; that is the rows and columns are interchanged. Therefore, the coordinates in the rotated coordinate system are written as the transpose of the rotation transformation matrix as $Y = X R^T$; Therefore, the covariance C' in the rotated coordinate system is:

$$C' = Y^T Y = \left[XR^T\right]^T \left[XR^T\right] = \left[R X^T\right]\left[XR^T\right] = R\left[X^T X\right]R^T = R\,C\,R^T \qquad \text{Eq 7.5}$$

The new covariance matrix is obtained by transforming the original covariance matrix using the matrix operation just given.

Exercise 7.4

Consider the covariance matrix in Exercise 7.1. Note that the non-diagonal elements are zero.

$$C = \begin{bmatrix} 0.035 & 0 \\ 0 & 0.011 \end{bmatrix}$$

The coordinate system is rotated using the following rotational transformation matrix:

$$R = \begin{bmatrix} 0.707 & -0.707 \\ 0.707 & 0.707 \end{bmatrix}$$

Calculate the covariance matrix in the new coordinate system. Interpret the results.

Answer

Using Eq 7.5, the covariance matrix in the rotated coordinate system is:

$$C' = \begin{bmatrix} 0.023 & 0.012 \\ 0.012 & 0.023 \end{bmatrix}$$

The given R matrix corresponds to a rotation of the coordinate system by 45°. In the original coordinate system, the variables x1, x2 are linearly independent. This is indicated by the zero values for the non-diagonal terms of C. In the rotated coordinate system, the new variables y1 and y2 are a linear combination of the original variables. They are not linearly independent. The covariance of y1 and y2 is non-zero, it is 0.012.

Another observation is that the maximum variance has decreased in the new coordinate system. The maximum value in C is 0.035, whereas it is 0.023 in C'. In general, maximum variance occurs in a coordinate system that is linearly independent. Any rotation of this coordinate system will result in lower values for variance, which can be intuitively explained by the product of sin and cos terms in the resultant matrix.

7.3.1 Singular value decomposition

If the variables in a problem are correlated, we can identify a new coordinate system in which the variables are linearly independent. We must find new variables that are not correlated. These variables are linear combinations of the original variables representing a transformation such as rotation or scaling. These variables represent directions in which the variance is the maximum. These directions are nothing but the eigenvectors of the covariance matrix.

For simplicity, we assume that the data matrix X is already centered (by subtracting values of each variable from its mean). Then the covariance matrix is written as $C = X^T X$. The eigenvalue equation is,

$$Cv = \lambda v \qquad \textit{Eq 7.6}$$

where, λ is the eigenvalue and v is the eigenvector. The significance of the eigenvector can be understood from Eq 7.6: when you multiply the matrix C to the eigenvector, you get the same vector scaled by a constant value λ. That is, the matrix C represents the scaling transformation of the eigenvector.

In Eq 7.6, λ is written as a diagonal matrix $[\lambda\ I]$, where I is the identity matrix consisting of one along the diagonals and zero everywhere else. The equation is rewritten as,

$$[C - \lambda I]\ v = 0 \qquad \textit{Eq 7.7}$$

The matrix $[C - \lambda I]$ contains the same elements as C except at the diagonals. λ is subtracted from the diagonal elements of the matrix C. Eq 7.7 represents a system of linear simultaneous equations in n variables. The unknown variables in the equations are the coefficients of the eigenvector v. Since the right-hand side is zero, the equations are satisfied only if the determinant of the matrix $[C - \lambda I]$ is zero. Setting the determinant to zero results in a polynomial equation of degree n, where $n \times n$ is the size of the matrix C. This equation is known as the *characteristic equation*. Solving the equation gives n values for λ, which are the eigenvalues. Let these be ordered in descending order as $(\lambda_1, \lambda_2, .., \lambda_n)$. Corresponding to each eigenvalue, an eigenvector can be found. Since the determinant is zero for each eigenvalue, substituting a value of λ in Eq 7.7 we get a system of linear simultaneous equations that are not independent. Therefore, we have the freedom to assume a value for one of the components of the eigenvector and solve for the remaining components. Finally, we divide all the components by the norm of the vector so that we get a unit vector. The following numerical example illustrates the procedure.

Exercise 7.5

The covariance matrix in the rotated coordinate system in Exercise 7.4 is:

$$C = \begin{bmatrix} 0.023 & 0.012 \\ 0.012 & 0.023 \end{bmatrix}$$

Calculate the eigenvalues and eigenvectors of this matrix.

Answer

The characteristic equation is obtained as:

$$|C - \lambda I| = 0$$

$$\begin{vmatrix} 0.023 - \lambda & 0.012 \\ 0.012 & 0.023 - \lambda \end{vmatrix} = 0$$

Expanding the expression for the determinant,

$$\lambda^2 - 0.046\lambda + 0.000385 = 0$$

Solving the quadratic equation, the eigenvalues are obtained as:

λ_1=0.035

λ_2=0.011

It can be verified that the eigenvalues are the same as the variances in the un-rotated coordinate system.

The eigenvectors are calculated by noting that one equation in the expression Eq 7.7 is redundant. Therefore, we can arbitrarily choose one component of the eigenvector as 1. Let v_{11}, v_{12} be the components of the first eigenvector. It should satisfy the expression:

$$(0.023 - \lambda_1)v_{11} + 0.012\, v_{12} = 0$$

Substituting for the first eigenvalue and the value of $v_{11} = 1$, we get $v_{12} = 1$. Both components of the first eigenvector are equal to 1, that is, the vector is rotated in the direction inclined at 45° to the original coordinate axis. Normalizing the eigenvector to get unit length, each component will be equal to cos(45). The second eigenvector is similarly obtained as (1, −1). It can be seen that, by taking the dot product of the two eigenvectors, we get 0. That is, the eigenvectors are orthogonal.

Some results

If $v_1, v_2, \ldots, v_n$ are the eigenvectors of the matrix C, they can be written in the form of an $(n \times n)$ matrix V in which each column represents an eigenvector as follows:

$$V = [v_1 \quad v_2 \quad \ldots \quad v_n] = \begin{bmatrix} v_{11} & \cdots & v_{n1} \\ \vdots & \ddots & \vdots \\ v_{1n} & \cdots & v_{nn} \end{bmatrix} \qquad \textit{Eq 7.8}$$

The eigenvalues can be written as a diagonal matrix L as follows:

$$L = \begin{bmatrix} \lambda_1 & 0 & \ldots & 0 \\ 0 & \lambda_2 & & 0 \\ & & \ldots & \\ 0 & 0 & & \lambda_n \end{bmatrix} \qquad \textit{Eq 7.9}$$

Since each eigenvector satisfies Eq 7.6, It can be verified that all the eigenvalues and eigenvectors satisfy the following matrix equation

$$C\,V = V\,L \qquad \textit{Eq 7.10}$$

If the matrix C is symmetric, we can find a set of eigenvectors that are orthogonal to each other. Vectors are orthogonal if their dot product is zero. Consider two eigenvectors v_1 and v_2 corresponding to eigenvalues λ_1 and λ_2.

$$C\,v_1 = \lambda_1\, v_1 \qquad \textit{Eq 7.11}$$

$$C\,v_2 = \lambda_2\, v_2 \qquad \textit{Eq 7.12}$$

Taking the transpose of Eq 7.12:

$$v_2^T\, C = \lambda_2\, v_2^T \qquad \textit{Eq 7.13}$$

Multiplying v_1 to both sides:

$$v_2^T C v_1 = \lambda_2 v_2^T v_1 \quad \textit{Eq 7.14}$$

From Eq 7.12:

$$v_2^T \lambda_1 v_1 = \lambda_2 v_2^T v_1 \text{ or,}$$

$$(\lambda_1 - \lambda_2)\, v_2^T v_1 = 0 \quad \textit{Eq 7.15}$$

Eq 7.15 can be satisfied only if $v_2^T\, v_1$ is equal to zero or $\lambda_1 = \lambda_2$. The first condition proves that the two vectors are orthogonal. If $\lambda_1 = \lambda_2$, we can still arbitrarily choose v_2 to be a vector orthogonal to v_1 satisfying Eq 7.12. Due to the orthogonality of the eigenvectors, the following result holds:

$$C = \lambda_1 v_1 v_1^T + \lambda_2 v_2 v_2^T + \ldots + \lambda_n v_n v_n^T \quad \textit{Eq 7.16}$$

This is easy to prove. Add an unknown error matrix E to the right-hand side as follows:

$$C = \lambda_1 v_1 v_1^T + \lambda_2 v_2 v_2^T + \ldots + \lambda_n v_n v_n^T + E \quad \textit{Eq 7.17}$$

Multiply both sides of the equation by v_1, v_2, etc. one after the other. Multiplying by v_1, we get

$$C\, v_1 = \lambda_1 v_1 v_1^T v_1 + \lambda_2 v_2 v_2^T v_1 + \ldots + \lambda_n v_n v_n^T v_1 + E\, v_1$$

That is,

$$C\, v_1 = \lambda_1 v_1 v_1^T v_1 + 0 + E\, v_1 \quad \textit{Eq 7.18}$$

Only the first term and the error term on the right-hand side will remain; all others will be zero because of the orthogonality of the vectors. Since the eigenvectors are normalized to have unit length, we get

$$C\, v_1 = \lambda_1 v_1 v_1^T v_1 + E\, v_1 = \lambda_1 v_1 + E\, v_1 \quad \textit{Eq 7.19}$$

Substituting for $C\, v_1$ on the (left-hand side) using the definition of the eigenvector (Eq 7.6), we get

$$\lambda_1 v_1 = \lambda_1 v_1 + E\, v_1$$

Therefore,

$$E\, v_1 = 0 \quad \textit{Eq 7.20}$$

Similarly, we get

$$E\, v_2 = 0, \text{ etc.} \quad \textit{Eq 7.21}$$

Any arbitrary vector x_1 can be written as a linear combination of the basis vectors v_1, v_2, etc. Therefore, we can combine all the previous equations to write

$$E\,x_1 = 0$$

Since this equation must be satisfied for any vector x_1, the error matrix should be zero. Therefore, Eq 7.16 holds. These results are useful for understanding principal component analysis (Chapter 11).

7.4 Probability

When we are not sure of the answer, we talk about probability! If an astrologer says that you are likely to have a good day today, it means he has no idea what is going to happen; he is making a wild guess. He makes a statement that can never be proved wrong because in essence, he is saying something may nor may not happen! In engineering, we resort to similar tactics in a more subtle manner, using mathematical jargon! Equipment manufacturers can never state for sure that their products will never fail. There are so many uncertainties in the operating conditions and human behavior. Therefore, they specify the failure probability instead of saying that the equipment is unsafe. This way, they are legally protected, and users are not worried about using an unsafe piece of equipment.

The classical example used to illustrate probability is tossing a coin. If you toss a evenly balanced coin, there is equal chance that you get heads or tails. That is, the probability of getting heads is 0.5 (50%). Statistically, this means that if you toss 1,000 times, about 500 times you will get heads. This is referred to as the frequentist interpretation of probability. For events that can be repeated many times, this procedure can be used to estimate probability. However, in many tasks we are required to quantify uncertainty, and there is no possibility of conducting repeated experiments. For example, the probability of failure of a modern bridge is as low as 10^{-6}, that is one in one million. It means that the bridge is not 100% safe. There is a small chance that the bridge could fail due to reasons such as: an unexpected heavy load happens to come on the bridge; or an unusual cyclone hits the area; or the material undergoes unanticipated corrosion due to some special conditions. Even though we cannot perform one million experiments consisting of various combinations of input conditions to determine how many times the bridge fails, we accept probability as a theoretical concept to refer to the chance of failure. Of course, we can perform computer simulations and calculate what is the percentage of cases in which the bridge is seen to be failing. This requires accurate modeling of the probabilities of the events involved.

Tossing a coin is an example of a discrete event. You get two outcomes, either heads or tails. There are many problems where you get a finite set of outcomes. Then you can specify the probability of each outcome. There are other situations where the outcome is a continuous range of values. For example, when you test the strength of different cubes made of the same mix of concrete, you get different values. You get these variations because there are many uncertainties in the mixing, casting, and testing processes. Here the outcome, cube strength, is a continuous variable and not a discrete variable. In such cases, we use probability density. A simple way to understand probability density is using a histogram. Divide the range of values into a number of intervals, and count the number of points that lie within each interval (frequency). If you plot this as a bar chart, you get a histogram such as in Figure 7.2. If you divide the frequency count by the total number of points, you get the probability of getting a point in that interval. If this probability is divided by the width of the interval,

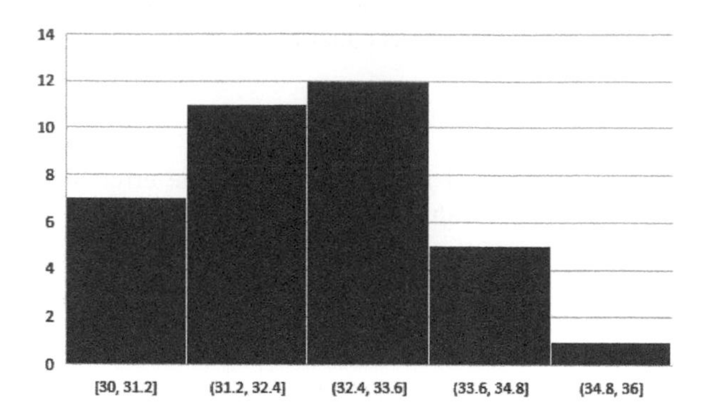

Figure 7.2 A histogram

you get the probability density. In other words, if you multiply the probability density by the interval width, you get the probability of getting a point in that interval. To use the terms of calculus, probability density is the derivative of probability. The cumulative probability is the integral of probability density. In Figure 7.3, the histogram is converted into a probability density function by fitting a smooth curve through all the discrete probability density values at the center of each interval. If you integrate the probability density function over the entire range, you get 1 since all the points lie within this interval.

Probability density functions of many naturally occurring random variables show standard forms. One common form is the normal distribution, also known as the Gaussian distribution. Another example is a binomial distribution (See Section 7.5.1.1), which represents the probability that a certain number of heads or tails is obtained when a coin is tossed several times.

7.4.1 Fundamental theorems of probability

The fundamental theorems of probability include the sum and product rules. The **sum rule** is written as,

$$P(A \cup B) = P(A) + P(B) - P(A \cap B) \quad \text{Eq 7.22}$$

P(A U B) is the probability of the union of the two events A and B, that is, the probability that either of the events occur. It is calculated as the sum of the probabilities of A and B minus the probability of both the events occurring simultaneously $P(A \cap B)$. The notation $A \cap B$ represents the intersection of both the sets.

Numerical Example: P(A) = 0.1 is the probability of an equipment failing, and P(B) = 0.2 is the probability of a worker not turning up for work. The work does not start on time if either event occurs. Then, the probability of work not starting on time P(C) = P(A) + P(B) = 0.3 (assuming that both events do not happen at the same time, that is, they are disjoint sets).

The **product rule** is written as follows:

$$P(A \cap B) = P(A) \cdot P(B|A) \quad \text{Eq 7.23}$$

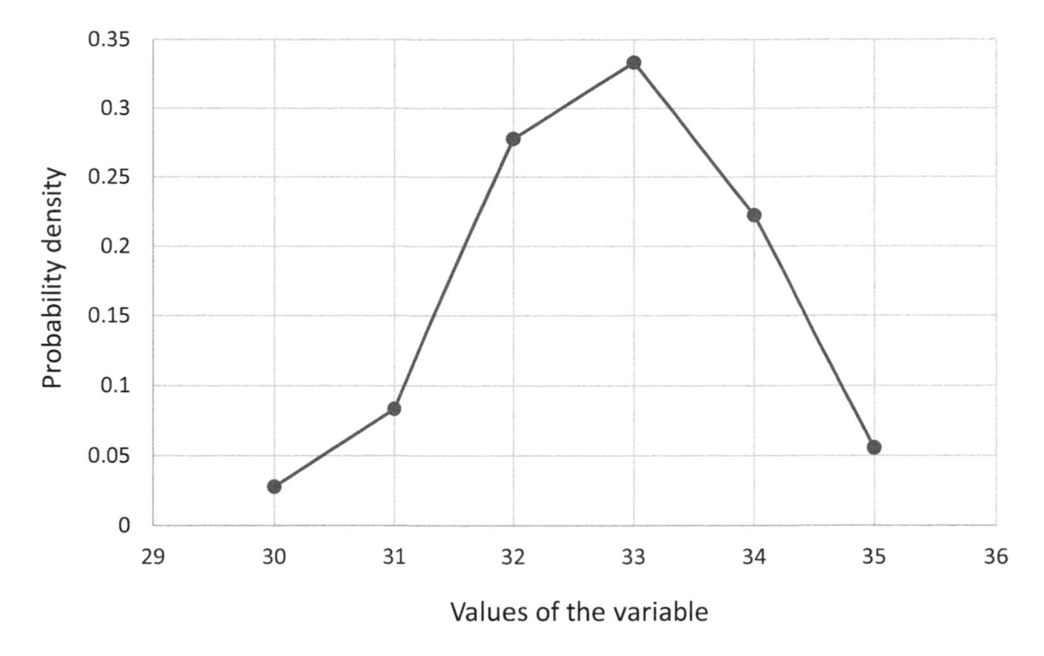

Figure 7.3 A probability density function

Here, P(A ∩ B) is the probability of both the events happening together. P(B|A) is the probability of B given A. That is, it is the probability of B happening after A has occurred. It is called the conditional probability of B given A. If A and B are independent events, that is, B is not influenced by A, then P(B|A) is the same as P(B). Then the probability of both the events happening is just the product of their individual probabilities.

Exercise 7.6

It has been observed that a piece of equipment breaks down once in ten days on average. The engineer who oversees repair of the equipment comes only on alternate days. What is the probability that the equipment breaks down on a day that the engineer is absent?

P(A) = 0.1 is the probability of the equipment failing. P(B) = 0.5 is the probability that the engineer is absent on any day. Therefore, the probability of both the events happening together = 0.1 * 0.5 = 0.05.

Exercise 7.7

Students can easily understand this simple example that explains the sum and product rules. Students are frequently faced with the dilemma – which topic they should spend more attention to while preparing for exams? A teacher set an online assignment in which students could choose one topic from Mechanics (M) or Strength of Materials (S). Once the topic is chosen, students cannot go back and change their decision. After choosing the topic, students are randomly assigned a question from the question bank for that topic. The teacher did not take care to ensure that all the questions have the same level of difficulty; some were difficult (D), some were easy (E). (Let us acknowledge that not all teachers are good and sincere.) The scenario is illustrated in Table 7.4. The two columns represent the question banks for M and S. M contains 60D and 40E. S contains 15D and 45E. Students do not know this distribution of questions in the question bank. Since one full day was given to select the assignment, students could share their questions with others who had not yet selected the assignment. How will they use this information to select an easy topic for the exam? Simply selecting the topic that is found to have more easy questions does not work because this might have happened by luck, especially when there is a limited amount of data.

If the data in Table 7.4 is available, the conditional probabilities can easily be calculated. $P(D|M) = 0.6$ and $P(D|S) = 0.25$. That is, after M is selected, the probability of getting a difficult question is $60/(60 + 40)$. Similarly, after S is selected, the probability of getting a difficult question is 15/60.

Initially, students had no idea about the conditional probabilities. So, they randomly selected the topics. Some got difficult questions and some got easy ones. The distribution of questions they got after one hour of this process is shown in Table 7.5. Note that for small sample sizes, the frequencies need not accurately match the theoretical probabilities. Also, the counts for M and S are not equal because students had their preferences and favorite topics.

Suppose 75% of students prefer S. What is the net probability that students get difficult questions? This is computed using the product rule. The probability of selecting S is 0.75. Therefore, the probability of selecting S and then getting a difficult question is $0.75 \times 0.4 = 0.3$. Similarly, the probability of getting a difficult question after choosing M is $0.25 \times 0.25 = 0.0625$. Therefore, the total probability of getting a difficult question is $0.3 + 0.0625 = 0.3625$.

Exercises based on Exercise 7.7

1 Calculate the net probability of getting an easy question using the product and sum rules. Calculate this for two cases, $P(S) = 0.75$ and $P(S) = 0.25$.
2 Calculate the sum of the probabilities of easy and difficult questions obtained in question 1. Verify that this sum is equal to 1.
3 Calculate the probability of easy questions using the total count in the second row of Table 7.4. Is the value obtained the same as that in question 1? What is the reason?
4 If someone got a difficult question, what is the probability that it is from question bank M? Use the numbers in Table 7.5 to do this calculation. Later, this can be compared with the value obtained using Bayes' theorem (discussed in the next section).

Table 7.4 Distribution of Easy and Difficult Questions in Two Question Banks

Difficult (D)	60	15
Easy (E)	40	45
	Mechanics (M)	Strength of Materials (S)

Table 7.5 Distribution of Randomly Selected Questions by Students

D (Total 10)	5	2
E (Total 11)	4	5
	M (Total 9)	S (Total 7)

Exercise 7.8

Another example similar to Exercise 7.7 is given here as a practice exercise. Two teachers, A and B, are teaching the same course to two batches of students. Students have no choice; they are randomly assigned their batches. Teacher A is known to give tough assignments and is more stringent in giving marks. (You can guess which teacher is more popular among students and what will happen if the students are not randomly allocated to the batches). Teacher A prefers to teach large classes; 100 students are assigned to his batch. Teacher B has only 40 students in his batch. From the performance of former students, it is seen that 20% of students of A fail the course, and 10% of students of B fail the course.

Questions

1. What is the probability that a student fails in the course this year?
2. What is the probability that a student who has failed in the course, belongs to batch A?
3. Suppose the head of the department decides to change the cut-off for the passing mark such that not more than 15% of students fail from the entire class. How will he do this? (This is an open-ended question.)

7.5 Bayes' theorem

Bayes' theorem follows directly from the product rule. It is used to update the probability of an event after an observation. It is written as:

$$\mathrm{P}(\mathrm{B} \mid \mathrm{A}) = \frac{\mathrm{P}(\mathrm{B})\mathrm{P}(\mathrm{A} \mid \mathrm{B})}{P(A)} \qquad \textit{Eq 7.24}$$

The left-hand side is the posterior probability of B given the fact that A has occurred. It is the conditional probability of B given A. P(B) is the prior probability of B, that is, the probability of B without knowing whether A has occurred or not. The theorem states that the posterior probability is proportional to the prior probability and to the conditional probability of A given B. Using Bayes' theorem, the conditional probability P(B|A) is computed from its inverse, that is, P(A|B). The denominator is the normalization factor that ensures that the total probability is 1. It is calculated using the formula:

$$\mathrm{P}(\mathrm{A}) = \sum\nolimits_{Y} \mathrm{P}(\mathrm{Y})\mathrm{P}(\mathrm{A} \mid \mathrm{Y}) \qquad \textit{Eq 7.25}$$

Here the probability of A is calculated as the sum of the conditional probabilities of A with the respect to all the possible events Y. The conditional probability of each event Y is the numerator of Eq 7.24. That is, the numerator of this equation is calculated for all the events, and then the sum is calculated, which is the normalizing factor used in the denominator.

The Bayesian probability updating might be performed like this. Suppose a number of observations are made in which the occurrence of the events A and B are recorded. Count the number of instances in which both A and B have occurred. The quantity P(A|B) on the right-hand side can be estimated by dividing this count by the total number of observations involving B. Using this, the numerator of Eq 7.24 is computed by multiplying by the prior probability P(B). Calculating the denominator requires enumerating all the possible events other than B.

For each event, the conditional probabilities need to be computed. Then, the posterior probability of B, P(B|A), can be calculated using Bayes' theorem. An example will make this clear.

Exercise 7.9 Price of concrete and steel

It is assumed that the probability of increase in the price of concrete (B) in the next month is 0.5. This is the prior probability based on certain beliefs of the user. Then it is noticed that the price of steel has started increasing (A). Should we update the probability of increase in the price of concrete?

Recent historical data are examined, and the number of times the prices of the two materials changed together is shown in Table 7.6.

Five out of six times when the price of concrete increased, the price of steel also increased. Therefore, the conditional probability of increase in steel price P(A|B) is obtained from this table as:

$$P(A|B) = (5/6)$$

The other event that could occur is A2, the price of steel decreases. The conditional probability of this is:

$$P(A2 \mid B) = (1/6)$$

Now, we need to compute the inverse of the conditional probabilities. From Bayes' theorem, the conditional probabilities of the two events B and B2 (increase and decrease in the price of concrete) are:

$$P(B \mid A) = \frac{P(B)\ P(A|B)}{P(A)} = \frac{0.5 \times \left(\frac{5}{6}\right)}{P(A)}$$

$$P(B2 \mid A) = \frac{P(B2)\ P(A|B2)}{P(A)} = \frac{0.5 \times \left(\frac{2}{4}\right)}{P(A)}$$

Note that P(B2) is taken as 1 – P(B) = 0.5. From Eq 7.25,

$$P(A) = \sum_{Y} P(Y)P(A|Y) = P(B)P(A|B) + P(B2)P(A|B2)$$
$$= 0.5 \times \left(\frac{5}{6}\right) + 0.5 \times \left(\frac{2}{4}\right) = \left(\frac{2}{3}\right)$$

Table 7.6 The Number of Times the Prices of Concrete and Steel Varied Together

	Concrete price decreased (B2)	*Concrete price Increased (B)*
Steel price decreased (A2)	2	1
Steel price increased (A)	2	5

Substituting in the denominator,

$$P(B \mid A) = \frac{P(B)\,P(A|B)}{P(A)} = \frac{0.5 \times \left(\frac{5}{6}\right)}{\left(\frac{2}{3}\right)} = 0.625$$

The Bayes' formula resulted in an increase in the estimate of probability of B from 0.5 to 0.625. This is based on historic observations as well as prior beliefs. Without considering prior probability, a naïve analysis would give a probability of increase in the price of concrete as (5/7) = 0.71, that is, the number of times concrete prices have historically increased when the price of steel has increased. This is not accurate because this estimate uses a limited amount of historic data that could be influenced by random effects. Bayesian updating using prior probability is a more scientifically accurate method.

7.5.1.1 Numerical example: application of Bayes' theorem – question bank

To illustrate Bayes' theorem, let us revisit the question bank example in Exercise 7.6. Students initially assumed that the distribution of questions in both topics is uniform, that is, equal percentage of easy and difficult questions are present in each question bank. Having observed previous students, they suspect that their assumption is not correct. They use Bayes' theorem to update the probabilities. To start with, let us consider only the topic S. Similar updating could be done for M as well. When you restrict your analysis to a single topic (S), the problem is similar to tossing a coin.

To use Bayes' theorem, you need a prior probability distribution. It is not enough to specify a single value such as 0.5 to indicate your assumption of the relative count of difficult and easy questions in the question bank. Instead, you need a density function that specifies the probabilities of different possible values of the variable. The discrete outcome of getting a difficult or easy question is a function of the unknown variable, the ratio of difficult and easy questions (R). The prior probability distribution of R tends to be subjective, but we can select a function that is most rationally justified based on our beliefs. In the absence of any other information, it makes sense to assume that easy and difficult questions are equally likely on average. That is, the R value of 0.5 should have the maximum probability. It is unlikely that all the questions in the question bank are either easy or difficult. So, the probability of getting these extreme ratios is 0. Now, the simplest option to define a function is to connect the maximum and minimum probability values with straight lines. This gives a triangular distribution for probability as shown in the bottom curve of Figure 7.4 (labeled as "prior"). This will be our prior probability distribution for Bayesian updating. To simplify the calculations, we will take only five possible values for R (0, 0.25, 0.5, 0.75, 1). Their corresponding probabilities are specified as (0, 0.25, 0.5, 0.25, 0), such that the values at the centroid of the intervals fit the triangular shape. Here, we have converted the continuous distribution into discrete values such that the total probability is 1 (Figure 7.4).

The assumed prior probabilities are updated when there is more data in the form of observations. We compute the posterior distribution based on the observations (O) as follows.

$$P(R|O) = P(R)\,P(O|R)/N \qquad \textit{Eq 7.26}$$

P(R) is prior probability of any selected value for the ratio. P(O|R) is the probability of getting the observed data using the specified ratio. It is known as the likelihood function. N is the normalization factor that ensures that the total probability is 1.0. The normalization factor is obtained by summing P(R|O) over all the values of R as described in the following.

Consider the value, R = 0.5, having prior probability P(R) = 0.5. In Table 7.5, after 7 trials in S, 2D and 5E were obtained. Without using Bayes' theorem, the ratio we get through a naive application of statistics using the sample is 2/7 = 0.286. But we do not know whether the computed ratio is accurate enough, because the sample size is small. Let us see what we get from Bayesian updating. This requires computing the likelihood function P(O|R). What is the probability of getting 2D in 7 trials if R = 0.5?

If R = 0.5, there is equal chance of getting easy and difficult questions. If the sample size is large, 50% of questions are likely to be difficult. That is the maximum probability should be at the ratio 0.5. Still, there is a finite probability that any ratio is obtained between 0 and 1. To get a ratio of 2/7, different sequences consisting of difficult and easy questions are possible as listed:

DDEEEEE, DEDEEEE, DEEDEEE, . . . , EEEEEDD

From your knowledge of permutations and combinations, you will realize that the number of ways in which you can get 2D and 5E is the number of combinations of 7 objects taken 2 at a time, written as $7C_2$. Its value is computed as:

$$7C_2 = (7 \times 6)/2 = 21$$

The total number of sequences possible is 128 (2^7). This is because at each position in a sequence, there are two possibilities D or E, and there are 7 such positions. Therefore, the number of possible combinations is $(2 \times 2 \times 2 \times 2 \times 2 \times 2 \times 2) = 2^7$. The probability of obtaining 2 difficult questions out of 7, P(O = 2D), is calculated as the fraction obtained by dividing the number of sequences with 2D by the total number of possibilities. That is:

$$P(O=2D|R = 0.5) = 21/128 = 0.164$$

In general, the probability of getting k difficult questions out of n trials (when the ratio is 0.5) is:

$$P_{k,n} = nC_k \times \left(\frac{1}{2}\right)^n \qquad \text{Eq 7.27}$$

Now, we can calculate the posterior probability of R = 0.5 from the observations. From Eq 7.26,

$$\begin{aligned} P(R = 0.5|O) &= P(R = 0.5)\ P(O|R = 0.5)/N \\ &= 0.5 \times 0.164/N \end{aligned}$$

The normalization factor N is calculated by summing up all the conditional probabilities. This will be done in the end.

Now we compute the posterior probability for R = 0.25. For this ratio, the probability of getting a difficult question P(D) = 0.25 and that for easy question is 1 – 0.25 = 0.75. In general, when there are two independent outcomes, if p is the probability of the first outcome, (1 – p) is the probability of the second outcome and the total probability of getting k instances of the first outcome out of n trials is given by:

$$P_{k,n} = nC_k p^k (1-p)^{(n-k)} \qquad \textit{Eq 7.28}$$

This function is called the Bernoulli or binomial probability. The expression is understood as follows. The probability of getting a sequence of k values of the first outcome is p^k and the probability of getting the remaining values of the second outcome is $(1 - p)^{(n - k)}$. Therefore, the probability of getting the sequence is the product of the two. There are nC_k ways of getting such a sequence, hence the formula Eq 7.28. Comparing this formula with Eq 7.27, it is seen that the probabilities, p and (1 – p), are substituted with the value (1/2) in Eq 7.28.

Substitute k = 2 and n = 7 in Equation 6.3.2, for the observation O = 2D,

$$P(O|R = 0.25) = 0.3115$$

The probability of getting the observed sequence is higher for R = 0.25, compared to R = 0.5. Repeating the calculations for R = 0.75,

$$P(O|R=0.75) = 0.0115$$

For R = 0 and R = 1, the likelihood function values are 0 because p = 0 or (1 – p) = 0. Using these values, the conditional probabilities are calculated using Eq 7.26. The results are tabulated in the second row in Table 7.7. In the third row the prior probabilities are multiplied with the likelihood function. The sum of the third row is the P(O) the probability of the given observation for all possible values of R considered here. P(O) is calculated to be 0.1626. This is the normalization factor. Dividing by this value, the posterior probability is calculated in the last row.

The prior and posterior probabilities are plotted in Figure 7.4. The posterior probability distribution has become skewed with higher probabilities on the left side. The maximum probability is still at R = 0.5, because the observations do not produce strong evidence to reverse the prior belief.

Exercise 7.10

In the previous example, calculate the posterior probability distribution if 2D and 6E were obtained in 8 trials.

Table 7.7 Conditional Probabilities

	R = 0	*R = 0.25*	*R = 0.5*	*R = 0.75*	*R = 1*
P(R)	0	0.25	0.5	0.25	0
P(O|R)	0	0.311	0.164	0.0115	0
P(R) * P(O|R)	0	0.0775	0.082	0.00288	0
P(R|O)	0	0.478	0.504	0.0176	0

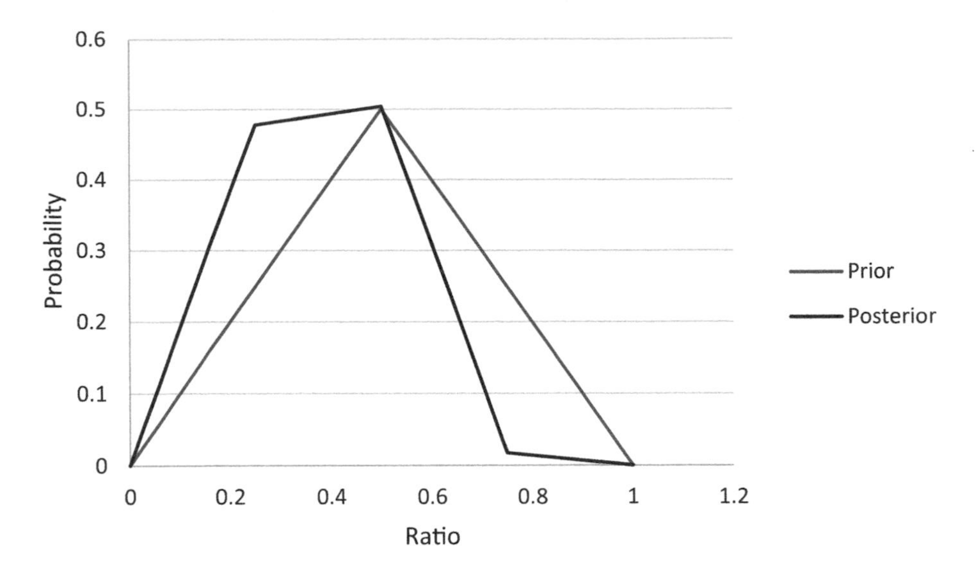

Figure 7.4 Prior and posterior probabilities

Answer

The posterior probability P(R = 0.25|O) becomes 0.583, higher than P(R = 0.5|O) = 0.41. With one additional observation compared to the previous example, it becomes clear that the ratio of 0.25 is more likely than 0.5. The probabilities are gradually updated using more evidence. Bayesian updating avoids rapid reversal of prior beliefs!

Exercise 7.11

In the same example, calculate the posterior probability distribution if 2E and no D were obtained in the first two trials. Discuss how the inference from the Bayesian updating procedure differs from the most likely estimate using simple statistics.

7.5.2 An example in fault diagnosis: tower crane

Consider this scenario inspired from real construction sites but simplified considerably for aiding the understanding of fundamental mathematical concepts. A tower crane undergoes large deflections while lifting a load. The crane is schematically shown in Figure 7.5. In practice, many things can go wrong during the operation of a crane. However, for the purpose of this discussion only two causes are listed:

a The connection at the base is damaged, and there is a large rotation at the fixed end of the cantilever.
b Jib suspension cable has yielded.

Let us analyze this situation using Bayes' theorem. Assume that close observation at the site is not possible, and we have to rely purely on sensors that measure deflections.

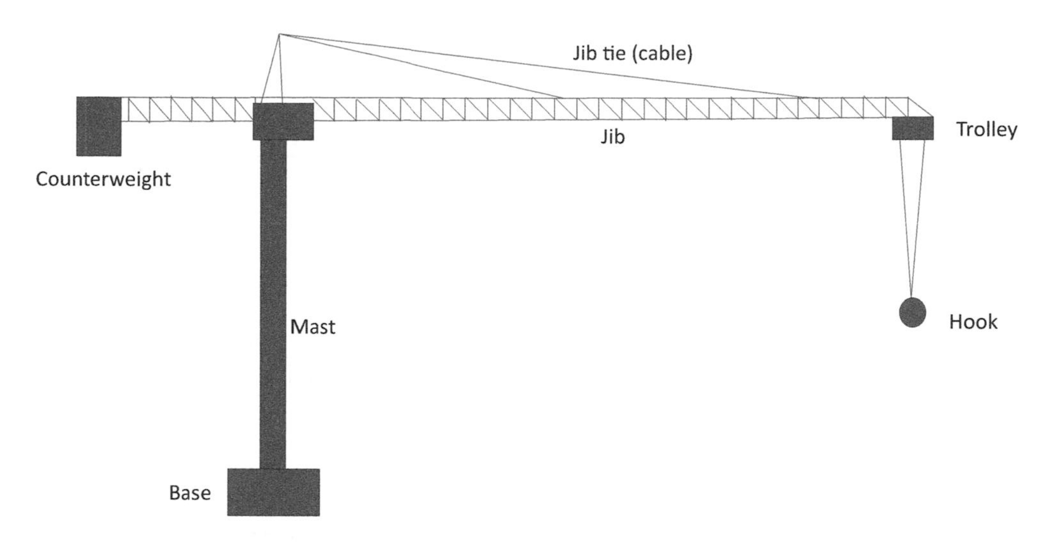

Figure 7.5 Schematic of a tower crane

Let the probabilities of the two causes listed in a and b be P(x = A) and P(x = B), where the variable x denotes the cause. Let y denote the effect, that is, deflections. The effect variable is discretized to represent only two values, large and small. Thus, we can write their probabilities as P(y = large) and P(y = small). Now, we need to compute the likelihood function, which involves conditional probabilities. What is the probability that the deflection is large, if the connection at the base is damaged (that is, under the condition x = A)? This is written as the conditional probability P(y = large | x = A), that is, the probability of large deflection, given the fact that the connection is damaged. As a short form, the variables x and y are removed from the bracket and the conditional probability is written as P(large|A). Similarly, the probability of large deflection under the condition that the jib suspension cable has yielded is written as P(large | B). These probabilities are tabulated in Table 7.8. Here, the columns represent the conditions y = large and y = small. The rows represent their causes x = A and x = B.

P(large | A) might be estimated using forward reasoning (simulations). For various degrees of damage possible at the support, structural analysis could be performed, and the number of situations that give large deflections could be counted to calculate the probabilities. The probability is computed as the fraction of such cases. Similarly, P(large | B) is computed by simulating the yielding of the cable for known lifting load. Let the computed probabilities be as follows:

P(large | A) = 0.9
P(large | B) = 0.2

The a-posteriori probabilities are computed using Bayes' theorem as follows:

$$P(A|\text{large}) = \frac{P(A) \times P(\text{large}|A)}{P(large)} \qquad \textit{Eq 7.1}$$

Table 7.8 Probability of deflection

	Large Deflection (y=large)	*Normal Deflection (y = small)*
Base damaged (x = A)	0.9	0.1
Jib tie yielded (x = B)	0.2	0.8

$$P(B \mid \text{large}) = \frac{P(B) \times P(\text{large} \mid B)}{P(large)} \qquad \textit{Eq 7.1}$$

Application of Bayes' theorem requires a-priori probabilities of events, P(A) and P(B). In the absence of additional information, we can assume equal probabilities of failure. That is,

P(A) = 0.5; and
P(B) = 0.5

The denominator in equations 7.1 and 7.2 is the normalization factor, which ensures that the total probability is 1. The probability that the deflection is large is the sum of the probabilities of large deflections under the two conditions A and B, which are the numerators in equations 7.1 and 7.2. Therefore, the a-posteriori probabilities in this case will be:

$$P(A \mid \text{large}) = \frac{P(A) \times P(\text{large} \mid A)}{P(large)} = \frac{0.5 \times 0.9}{0.5 \times 0.9 + 0.5 \times 0.2} = 0.818$$

$$P(B \mid \text{large}) = \frac{P(B) \times P(\text{large} \mid B)}{P(large)} = \frac{0.5 \times 0.2}{0.5 \times 0.9 + 0.5 \times 0.2} = 0.182$$

In this case, the large deflection is most likely due to the damage at the support because the probability of large deflection due to this cause is higher than that due to the yielding of the cable. The a-posteriori probabilities were mostly influenced by the relative values of P(large | A) and P(large | B) since the a-priori probabilities were equal. However, if additional information is available, the probabilities of A and B can be estimated more accurately. For example, if routine inspections of the connections are done systematically and processes are in place to achieve a target reliability of 99.99%, the probability of A might be reduced to 0.01. Then the Bayes' theorem reveals there is a higher probability that the high deflection is due to the yielding of the cables as shown in the following calculations:

$$P(A \mid \text{large}) = \frac{P(A) \times P(\text{large} \mid A)}{P(large)} = \frac{0.01 \times 0.9}{0.01 \times 0.9 + 0.5 \times 0.2} = 0.0826$$

$$P(B \mid \text{large}) = \frac{P(B) \times P(\text{large} \mid B)}{P(large)} = \frac{0.5 \times 0.2}{0.01 \times 0.9 + 0.5 \times 0.2} = 0.9174$$

In this case, the likely cause of large deflection will be identified as the yielding of the cable.

7.5.3 Bayes' theorem and machine learning

As mentioned before, many machine learning tasks are inverse problems. For solving an inverse problem, the Bayesian updating formula can be explained like this:

The probability of a cause is updated by multiplying a) the estimate of prior probability of the cause and b) the likelihood of the observed effects, which is normalized by dividing by the total probability of the observed effects due to all possible causes.

Bayes' theorem plays a crucial role in machine learning. Most machine learning techniques use statistical methods for finding the right hypothesis that explains data. Their performance can be studied from a Bayesian viewpoint. In general, the posterior probability of a hypothesis given a data set is computed from an assumption of prior probabilities. Even though, this book does not include Bayesian interpretation of machine learning, it is pointed out that this is an alternative viewpoint that provides interesting insights into the working of the algorithms. Readers are referred to Bishop (2006) for a detailed treatment of such methods.

7.6 Optimization

Most machine learning tasks are formulated as optimization problems. For a quick introduction to optimization, see Chapter 8 of the book Raphael and Smith (2013). Key concepts are discussed here.

Decision variables (Optimization variables): These are the variables for which we need to find the best combination of values such that our objective is achieved.

Objective function: This is the function that is minimized or maximized. Since a maximization is possible by simply taking the negative of the objective function that is minimized, we often talk about minimizing an objective function, without loss of generality. The objective function should be written in terms of the decision variables. Sometimes, you might have an explicit mathematical expression for the objective function containing the decision variables. However, in most engineering problems, it is difficult to write an explicit mathematical expression. A common strategy is to generate a simulation model using the given values of optimization variables. By simulating the model, the output parameters are obtained. The objective function is written in terms of the output parameters obtained through simulation.

Constraints: In addition to minimizing the objective function, we also need to satisfy the constraints of the problem. Constraints can take different forms, equalities, inequalities, linear, non-linear, etc. The set of solution points that satisfy all the constraints is called the feasible space. We are interested in the minimum (or maximum) value of the objective function within the feasible space.

Optimization algorithms: The fundamental mathematical technique for finding the optimum of a function is to set the derivatives of the objective function to zero with respect to all the decision variables. This is the necessary condition for obtaining a maximum or minimum of a function; the slope is zero at these points. When the objective function is simple and written in as an explicit mathematical equation, the derivatives can be computed analytically.

However, for many practical problems this is not possible. Exact mathematical techniques are often inadequate to deal with the complexities of practical engineering tasks. Therefore, numerical methods are typically used for solving optimization problems. A simple method is to randomly generate many combinations of values of variables. The solution point that gives the minimum value of the objective function is taken as an approximation to the solution. Since this procedure does not converge fast enough, other random search methods have been developed, making use of meta-heuristics to generate solution points in regions where there is faster improvement in the value of the objective function. An example of such method is PGSL (See Chapter 8 of Raphael and Smith, 2013).

Application to control tasks: To determine the best control action for maximizing our objectives, we need to develop an optimization model. This involves identifying the variables, their characteristics, ranges of values, etc. The objective function should also be defined in terms of the decision variables. When the parameters that influence the objectives can only be computed through simulation programs, the objective function is defined as a "black-box", a function that computes the objective function for a given set of values of decision variables. We may not be able to write down explicit mathematical expression for the objective function. When random search methods are used, combinations of values of decision variables are randomly generated by the optimization algorithm; then the user-defined objective function is called to evaluate the solution. The process is repeated, and the best solution is reported at the end of the search process.

7.7 Summary

Essential mathematics required for understanding machine learning techniques is introduced in this chapter. First, the concept of an inverse problem is described. In an inverse problem, we attempt to infer the actions from their effects. Inverse problems are common in many automation and control tasks. Then fundamental concepts related to statistics and probability are briefly covered. The significance of covariance, correlation, and other statistical parameters in solving engineering problems is illustrated with examples. The application of Bayes' theorem is also explained using examples from engineering.

References

Bishop, C.M. (2006). *Pattern Recognition and Machine Learning*, New York, USA: Springer.

Papadopoulou, M., Raphael, B., Smith, I.F.C., and Sekhar, C. (2016). Optimal Sensor Placement for Time-Dependent Systems: Application to Wind Studies Around Buildings, *ASCE Journal of Computing in Civil Engineering*, 30 (2), pp. 04015024.

Raphael, B. and Smith, I.F.C. (2013). *Engineering Informatics: Fundamentals of Computer Aided Engineering*, second edition, Chichester, UK: John Wiley.

Vernay, D.G., Raphael, B., and Smith, I.F.C. (2014). Augmenting Simulations of Airflow Around Buildings Using Field Measurements, *Advanced Engineering Informatics*, 28 (4), pp. 412–424.

Vernay, D.G., Raphael, B., and Smith, I.F.C. (2015). Improving Simulation Predictions of Wind Around Buildings Using Measurements Through System Identification Techniques, *Building and Environment*, 94, pp. 620–631.

8 Regression

8.1 Introduction

Regression was introduced in Chapter 6. This chapter presents details of the mathematical models and algorithms for this task.

8.1.1 Example: predicting lighting in a room

Automated blinds are installed on windows in a room to control the amount of heat and light transmitted into the room. The blinds are raised or lowered using motors according to the amount of daylight available. The blinds are lowered when there is too much daylight in the room, causing glare. The blinds are fully raised when lighting levels are low. In many such applications, the actions are taken iteratively through closed-loop feedback control. This involves making small changes and observing its effects. The process is repeated until the objectives are met. However, this strategy may not be appropriate in certain cases. Too much movement of the blinds might be irritating to the user. If actions can be taken in a single step, it causes minimum distraction to the user. This requires predicting what will be the light level for a given blind position and outdoor conditions. Lighting simulation software such as Radiance (Ward and Shakespeare, 1998) have been used to predict the lighting inside buildings. Lighting simulation requires an accurate 3D model of the building along with material properties and weather data. Even if the model is available, simulations take considerable time, and it may not be appropriate for real-time control. How can predictions be made faster?

Surrogate models have been used for fast predictions. The concept involves developing a simple empirical model and calibrating it with data. The training data could be obtained by past simulations or actual measurements. Once the model is trained, it can predict the output variables for new values of the input variable.

In this example, let us take the input variables to be outdoor lighting level and the blind position. We want to predict the indoor lighting level at a point within the room. If many simulations are done for various combinations of outdoor conditions and blind positions, this data could be used for training. A sample data set is given in Table 8.1.

A simple surrogate model is $y = w1 \times x1 + w2 \times x2 + b$, where x1 and x2 are input variables (outdoor lighting and blind position), y is the output variable (indoor lighting), (w1, w2, b) are the model parameters to be determined from the data. This is a regression problem. We must find the model parameters such that the predictions by the model closely match the actual values given in the table. Once this model is developed, the lighting level can be predicted without performing a complex simulation.

DOI: 10.1201/9781003165620-10

Table 8.1 Indoor Lighting Levels for Various Blind Positions

Outdoor Lighting (lux) (x1)	*Blind Position (% opening) (x2)*	*Indoor Lighting (lux) (y)*
10,000	100	2000
8000	20	120
1000	50	80
...		
3000	80	?

8.2 Linear regression

In a linear model, the output variable is a linear combination of the input variables, written as,

$$y_p = \Sigma w_j x_j + b \qquad \text{Eq 8.1}$$

Where, y_p is the predicted value of the output variable, x_j is the j-th variable and w_j is the coefficient of the j-th variable. b is called the *bias* term, which has the effect of producing an output even when all the input values are zero. This equation is written using vector notation as

$$y_p = \langle \mathbf{w}, \mathbf{x} \rangle + b \qquad \text{Eq 8.2}$$

Here, **w** and **x** are vectors of dimension *n*, *n* is the number of input variables. Note that vectors are written in boldface. The vector **x** contains all the input variables x_j, and **w** contains the corresponding weight factors w_j. The angle bracket represents the dot product of these two vectors. This is computed by multiplying each weight factor w_j by the corresponding value of x_j and summing up all the products. This is a convenient short form of Eq 8.1. The coefficients w_j are determined such that the error between the predicted and actual values of the output variable is the minimum.

8.2.1 Geometric interpretation

Eq 8.2 represents a hyperplane in *n* dimensions, which is equivalent of a planar surface in three dimensions. In one dimension (one input variable), it is a straight line that passes through all the points as close as possible.

The vector **w** defines a direction perpendicular to the hyperplane. By changing the value of **w**, the hyperplane is rotated. The quantity $\langle \mathbf{w}, \mathbf{x}_i \rangle$ is the projection of the vector $\mathbf{x}_i$ on the normal vector **w**, that is, it is proportional to the distance of the point from the hyperplane. The parameter b determines how much the hyperplane is shifted from the origin. Varying the value of b moves the hyperplane parallel to itself. The error in the model prediction for a given point $\mathbf{x}_i$ is given by

$$e_i = y_i - \left(\langle \mathbf{w}, \mathbf{x}_i \rangle + b \right) \qquad \text{Eq 8.3}$$

This is also called the *loss*, since it represents the deviation from the real situation. The loss should be reduced to as low as possible by adjusting the values of **w** and b. This is done by optimization, by minimizing a suitable *loss function* that combines the errors in all the data points.

8.2.2 Least squares approach

In the least squares approach, the loss function is the sum of squares of errors in all the data points. It is written as

$$L(\boldsymbol{w},b)=\sum_{i=1}^{l}\left(\langle \boldsymbol{w},\boldsymbol{x_i}\rangle+b-y_i\right)^2 \qquad \textit{Eq 8.4}$$

To minimize this function, the derivatives with respect to all the optimization variables (decision variables) must be set to zero. Here, the optimization variables are **w** and b. Using the summation notation instead of the dot product of vectors $\langle \boldsymbol{w},\boldsymbol{x_i}\rangle$, the same function is written as follows:

$$L(\boldsymbol{w},b)=\sum_{i=1}^{l}\left(\sum_{j=1}^{n}\left(w_j\, x_{i,j}\right)+b-y_i\right)^2 \qquad \textit{Eq 8.5}$$

Note that the summation over *i* is for all the data points, and the summation over *j* is for all the input variables. $x_{i,j}$ is the value of the *j*-th variable in the *i*-th data point. Differentiating with respect to w_j,

$$\frac{\partial L}{dw_j}=2\sum_{i=1}^{l}\left(\sum_{j=1}^{n}\left(w_j\, x_{i,j}\right)+b-y_i\right)x_{i,j} \qquad \textit{Eq 8.6}$$

Noting that the term inside the first bracket is the error defined by Eq 8.3, and setting the derivative to zero,

$$\frac{\partial L}{dw_j}=2\sum_{i=1}^{l}e_i x_{i,j}=0;\rightarrow\sum_{i=1}^{l}e_i x_{i,j}=0 \qquad \textit{Eq 8.7}$$

With this result, we now revert to the vector notation. For mathematical convenience, we include the bias also in the weight vector as the last term, that is, [**w**,b] = [w_1, w_2, . . . , w_n, b]. Then we include a dummy last element in the input vector having 1 for all the data points, that is, [x_1, x_2, ... , x_n, 1]. With this, we can treat b in the same way as the weight factors. The prediction is now written as $y_p=\langle \boldsymbol{w},\boldsymbol{x}\rangle$ omitting the b term (because the dot product includes b as the last term).

Now, define an input data matrix **X** in which each row corresponds to one data point and each column corresponds to an input variable. Also, define an error vector ***e*** in which each row contains the error for each point, and an output data matrix Y similarly. Noting that Eq 8.7 is valid for each input variable j, and it involves the dot product of the transpose of the **X** and the error vector **e**, Eq 8.7 is rewritten as

$$\boldsymbol{X}^{\boldsymbol{T}}.\boldsymbol{e}=0 \qquad \textit{Eq 8.8}$$

That is, the error vector is orthogonal to each input data vector. The error vector is the difference between actual and prediction values, and can be written as follows

$$e = Y - X.w \qquad \text{Eq 8.9}$$

Multiplying $\mathbf{X}^T$ on both sides of the Eq 8.9,

$$X^T.e = X^T.Y - X^T.X.w \qquad \text{Eq 8.10}$$

Substituting Eq 8.8 in Eq 8.10,

$$0 = X^T.Y - X^T.X.w \text{; or}$$

$$\left[X^T X\right] w = X^T Y \qquad \text{Eq 8.11}$$

$\left[X^T X\right]$ is a square matrix of dimension $n \times n$, where n is the number of input variables. $X^T Y$ is a column vector of dimension n. Therefore, this equation represents a system of linear simultaneous equations in n variables, the unknown variables being the components of the **w** vector. By solving these equations, we get the weight vector that minimizes the loss function Eq 8.4. The equations can be solved by multiplying both sides by the inverse of the $\left[X^T X\right]$ matrix. After obtaining the weight vector, the output for a new test point can be predicted by taking the dot product of the two vectors.

Note that $\left[X^T X\right]$ is the same as the covariance matrix if the data are normalized with zero mean. The matrix cannot be inverted if its determinant is zero, that is, if the matrix is singular.

Exercise 8.1

The input and output variables for an application are given in Table 8.2. Perform linear regression. Qualitatively explain the results.

Table 8.2 Regression Data

x1	*y*
1	2
2	3
3	3

Answer

The problem has only one input variable. As explained in the previous section, the input vector is modified by adding 1 as the last element to each variable so that when multiplied with the modified weight vector [**w**,b] the expression for predicting the output contains the bias term correctly. Therefore, the X and Y matrices are written as follows

$$X = \begin{bmatrix} 1 & 1 \\ 2 & 1 \\ 3 & 1 \end{bmatrix}; Y = \begin{bmatrix} 2 \\ 3 \\ 3 \end{bmatrix}$$

Computing the terms in Eq 8.11

$$\left[X^T X\right] = \begin{bmatrix} 1 & 2 & 3 \\ 1 & 1 & 1 \end{bmatrix} \begin{bmatrix} 1 & 1 \\ 2 & 1 \\ 3 & 1 \end{bmatrix} = \begin{bmatrix} 14 & 6 \\ 6 & 3 \end{bmatrix}$$

$$\left[X^T Y\right] = \begin{bmatrix} 1 & 2 & 3 \\ 1 & 1 & 1 \end{bmatrix} \begin{bmatrix} 2 \\ 3 \\ 3 \end{bmatrix} = \begin{bmatrix} 17 \\ 8 \end{bmatrix}$$

Therefore, Eq 8.11 gives

$$\begin{bmatrix} 14 & 6 \\ 6 & 3 \end{bmatrix} \begin{bmatrix} w_1 \\ b \end{bmatrix} = \begin{bmatrix} 17 \\ 8 \end{bmatrix}$$

Solving by taking the matrix inverse,

$$\begin{bmatrix} w_1 \\ b \end{bmatrix} = \begin{bmatrix} 14 & 6 \\ 6 & 3 \end{bmatrix}^{-1} \begin{bmatrix} 17 \\ 8 \end{bmatrix} = \begin{bmatrix} 0.5 \\ 1.667 \end{bmatrix}$$

Table 8.3 Predicted Output Variables

x1	*y*	y_p	*Error*
1	2	2.167	−0.167
2	3	2.667	+0.333
3	3	3.167	−0.167

Using the values of **w**,b, the predicted output values for each point are given in Table 8.3.

It is noted that the predicted values differ from the actual values of y. This is because, there are two variables (w1 and b) and there are three data points. It is not possible to find values for w and b such that all the output values match exactly. It can be verified that, if the value of y in the last point were 4 (instead of 3), the values of [w,b] become [1,1] and the prediction errors are 0.

In Table 8.3, it is noticed that the errors are evenly distributed on the positive and negative side such that the net error is zero. The weights obtained by regression represent a mean hyperplane that passes through the set of points such that the errors are balanced on the positive and negative sides.

Exercise 8.2 Predicting the swing radius of a crane

When a crane lifts an object, it may swing sideways, especially when the load is eccentric to the hoist at the time of lifting. It is hazardous for the human operators and other people on site. Data are collected from a site to predict the swing radius of the crane and is presented in Table 8.4. There are two input variables, the speed of lifting, and the load. These are converted to normalized units (dimensionless) in the table. Is it possible to develop a prediction model through regression using this data? Explain.

Table 8.4 Crane Swing Radius with All Variables Normalized

Speed	*Weight*	*Swing Radius*
0.2	1.0	0.1
0.4	0.8	0.2
0.6	0.6	0.3
0.8	0.4	0.4
1.0	0.2	0.5

Answer

It is convenient to use a spreadsheet or a script to complete this exercise. The X matrix is given in the following:

Speed	*Weight*	*Bias Term*
0.2	1.0	1
0.4	0.8	1
0.6	0.6	1
0.8	0.4	1
1.0	0.2	1

Calculating $[\mathbf{X}^T \mathbf{X}]$

$$\left[X^T X\right] = \begin{bmatrix} 2.2 & 1.4 & 3 \\ 1.4 & 2.2 & 3 \\ 3 & 3 & 5 \end{bmatrix}$$

Solving for the weights requires taking the inverse of this matrix. However, the determinant of this matrix is 0, so the inverse does not exist! Therefore, linear regression is not possible for this data.

This happened because the two input variables are strongly correlated in Table 8.4. The variables are not independent; in the given data, load does not vary independently of the speed. Therefore, the effect of load on the output variable cannot be estimated from this data.

8.2.3 Autoregression

Autoregression is frequently used in predictions with time series. Sensors and other sources produce data at regular intervals. Frequently, the output at the next time step is correlated with the data in previous time steps. This typically happens when the value of the output variable is determined by differential equations involving time. Such differential equations can be converted to finite difference forms in which the output at time *t* can be written in terms of the values of variables at previous time steps (t − 1), (t − 2), etc. Thus, the data from previous time steps can be used to predict the value at the next time step. This can easily be converted into a regression problem. The regression equation is as follows:

$$y(t) = w_1\, y(t-1) + w_2\, y(t-2) + \ldots + b$$

The coefficients w_1, w_2 etc. are determined through regression. All that needs to be done is to arrange data in a form such that the columns representing input variables contain the data at previous time steps.

8.2.4 Difficulties with linear regression

When there is strong correlation between input variables, the equations become ill-conditioned. It may not be possible to invert the coefficient matrix because the determinant becomes close to zero. This condition is called ***multi-collinearity***. A singular covariance matrix means that one of the variables is superfluous – it can be represented as a linear combination of the others. To avoid this, a minimum number of variables should be carefully chosen in the model. Domain knowledge is important here. Chosen variables must be known to influence the output variables through physical principles or logical reasoning. The data set should also be carefully prepared. You must make sure that there is enough variability in the data; that is, each variable should vary independently.

Another difficulty is the high computational costs when dealing with large data sets. Minimizing the least squares loss function involves solving a set of linear simultaneous equations involving large number of variables. The X matrix is potentially large, and the matrix inversion is an expensive operation. Iterative procedures are sometimes preferred when there are many data points and matrix operations take too much time and memory.

8.3 Ridge regression

As the number of variables increases, it becomes easier to fit all the training data by adjusting the weight factors. There are many problems where the number of variables is much more than the number of training data points. Linear regression does not work in such cases. Either the coefficient matrix becomes singular or there is overfitting. *Overfitting* refers to the condition when the model predictions match all the data points because of the large number of model parameters that can be tuned; however, it produces wrong predictions in unseen test points. In such cases, it is better to find simple models that produce approximate results rather than complex models that have poor generalizability. One way to do it is to reduce the number of model parameters. Ridge regression does that.

In ridge regression, the norm of the weight vector is minimized along with the error function. Both are combined into a single loss function L as follows:

$$\mathrm{L}(\mathrm{w},\mathrm{b}) = \lambda \langle w, w \rangle + \sum_{i}^{l} \left(y_i - \left(\langle w, x_i \rangle + b \right) \right)^2 \qquad \textit{Eq 8.12}$$

The first term on the right-hand side represents the minimization of the norm of the weight function. The second term is the same as the square loss function in regression. λ is a parameter that controls the trade-off between square loss and norm of the weight vector. Eq 8.12 is called the penalized loss function because there is a penalty for the large norm of the weight vector. This promotes selection of solutions containing low or zero weights for many variables, thereby selecting the simplest hypothesis with minimum variables that can explain the data. Adding the term that reduces the complexity of the hypothesis is referred to as *regularization*. The regularizer selected here is known as *weight decay* or *parameter shrinkage* since it encourages the weights to decay towards zero unless it is necessary to reduce the prediction errors.

The loss function is minimized by setting the derivatives to zero, as was done in the case of linear regression. Since the method has already been demonstrated before, the derivation is skipped here. The final solution obtained is as follows:

$$\left[\boldsymbol{X}^T\boldsymbol{X} + \lambda I\right]\boldsymbol{w} = \boldsymbol{X}^T\boldsymbol{Y} \qquad \text{Eq 8.13}$$

where I is the unit square matrix containing only diagonal elements, each having the value one. Comparing with Eq 8.11, the only difference is that the quantity λI is added to the diagonal elements of $\left[\boldsymbol{X}^T\boldsymbol{X}\right]$. This has the effect of adding numerical stability to the equations by making the determinant non-zero. Even in cases where the covariance matrix is singular, Eq 8.13 gives valid solutions.

Exercise 8.3 Ridge regression – predicting the swing radius of a crane

Perform ridge regression using the data in Table 8.4 for $\lambda = 1$ and $\lambda = 0.01$. Explain the results.

Answer

Using $\lambda = 1$, $\left[\boldsymbol{X}^T\boldsymbol{X} + \lambda I\right]$ is

$$\begin{bmatrix} 2.2 & 1.4 & 3 \\ 1.4 & 2.2 & 3 \\ 3 & 3 & 5 \end{bmatrix} + \begin{bmatrix} 1 & 0 & 0 \\ 0 & 1 & 0 \\ 0 & 0 & 1 \end{bmatrix} = \begin{bmatrix} 3.2 & 1.4 & 3 \\ 1.4 & 3.2 & 3 \\ 3 & 3 & 6 \end{bmatrix}$$

The inverse of this matrix gives

$$\begin{bmatrix} 3.2 & 1.4 & 3 \\ 1.4 & 3.2 & 3 \\ 3 & 3 & 6 \end{bmatrix}^{-1} = \begin{bmatrix} 0.590 & 0.035 & -0.313 \\ 0.035 & 0.590 & -0.313 \\ -0.313 & -0.313 & 0.479 \end{bmatrix}$$

Multiplying the inverse of the matrix with $\boldsymbol{X}^T\boldsymbol{Y}$ gives

$$\begin{bmatrix} w_1 \\ w_2 \\ b \end{bmatrix} = \begin{bmatrix} 0.205 \\ -0.017 \\ 0.156 \end{bmatrix}$$

Using these values, the actual and predicted values are compared in the following table:

Speed	*Load*	*Actual Swing*	*Predicted Swing*	*Error*
0.2	1.0	0.1	0.18	−0.08
0.4	0.8	0.2	0.224	−0.024
0.6	0.6	0.3	0.269	0.031
0.8	0.4	0.4	0.313	0.087
1.0	0.2	0.5	0.358	0.142

Using the value of λ=0.01, the same calculations are repeated to obtain

$$\begin{bmatrix} w_1 \\ w_2 \\ b \end{bmatrix} = \begin{bmatrix} 0.351 \\ -0.142 \\ 0.174 \end{bmatrix}$$

Using these values, the actual and predicted values are compared in the following table:

Speed	*Load*	*Actual Swing*	*Predicted Swing*	*Error*
0.2	1.0	0.1	0.102	−0.002
0.4	0.8	0.2	0.201	−0.001
0.6	0.6	0.3	0.300	0.000
0.8	0.4	0.4	0.398	0.002
1.0	0.2	0.5	0.497	0.003

The errors in prediction for λ=0.01 are much lower than for λ=1. However, the weights are higher. In both cases, it is possible to find weight factors that reasonably predict the output. This is in contrast with the case of linear regression, where no solution was obtained because the matrix was singular.

Ridge regression eliminates some of the problems associated with least squares regression. Ill-conditioning of equations due to multi-collinearity in data are eliminated through the regularization parameter added to the diagonal of the covariance matrix.

8.4 Non-linear regression

In many applications, output variables depend on the input variables through non-linear relationships. Linear regression results in large errors in such cases, if a relationship of the form Eq 8.1 is learnt. In Figure 8.1, deflections in a concrete beam obtained in a laboratory experiment are plotted against the measured widths of a crack near the midspan (Stephen et al., 2019). Two curves are shown. The first one is the experimental data and the second one the predictions of a non-linear model that was calibrated using the data. It is clear that a linear relationship between the two variables will not be appropriate to model the behavior of this beam.

Prediction errors might be reduced by modeling non-linear relationships between input and output variables. A simple solution is to use higher-order polynomials in the model. For example, an output variable y that depends on two input variables x_1, x_2, might be modeled as follows

$$y = w_1x_1 + w_2x_2 + w_3x_1^2 + w_4x_2^2 + w_5x_1x_2 + b \qquad \text{Eq 8.14}$$

Here, quadratic terms have been added to the expression. As before, the weight factors must be determined by minimizing the prediction errors. It might not be obvious at first sight that this is still a linear learning problem. Even though there are quadratic terms involving the input variables, the relationship is linear with respect to the model parameters **w**. Therefore,

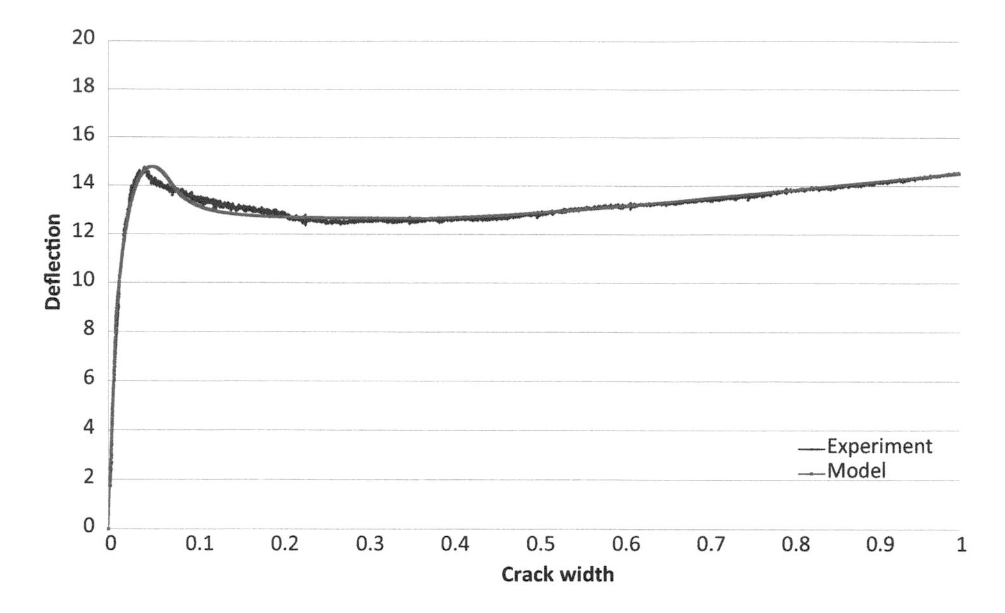

Figure 8.1 Deflection in a concrete beam obtained in an experiment

the same techniques that were discussed earlier could be applied. If we introduce new variables x_3, x_4 and x_5 as $x_3 = x_1^2; x_4 = x_2^2; x_5 = x_1 x_2$, Eq 8.14 is written as

$$y = w_1x_1 + w_2x_2 + w_3x_3 + w_4x_4 + w_5x_5 + b \qquad \textit{Eq 8.15}$$

This is a linear relationship of the form of Eq 8.1. The newly introduced variables are called features and are computed using functions of the original input variables.

Non-linear relationships between input and output variables could be learnt by using features, which are derived quantities computed using the original input variables. This technique is called learning in the feature space.

8.4.1 Learning in the feature space

A feature is a relationship between input variables. In general, it is a vector that is computed using the values of input variables. A scalar feature is written as

$$f_1 = F_1\,(x_1, x_2, \ldots, x_n),$$

where F_1 is a function of the input variables. For example, $f1 = x_1 \times x_2^2$ is a non-linear feature, a monomial of degree 3. If the *p* numbers of such features are defined, they can be written in vector form as:

$$\varnothing(\boldsymbol{x}) = \begin{bmatrix} f_1(\boldsymbol{x}) \\ f_2(\boldsymbol{x}) \\ \cdots \\ f_p(\boldsymbol{x}) \end{bmatrix}$$

Here, the function $\varnothing(x)$ maps the original variables (x_1, x_2, . . . , x_n) into the feature space. The functions used to compute the features are called the *basis* functions. The relationship between input and output variables is written as a linear combination of the basis functions.

The feature space is the set of all possible values of feature vectors, that is, it is a set in which each element is a vector of dimension *p*. A *p*-dimensional space consisting of *p* real-valued variables is denoted by the symbol $\mathbf{R}^p$. Note that the dimension *p* of the feature vector will be generally different from the dimension of the original input vector.

Exercise 8.4

A learning task involves three variables (x_1, x_2, x_3). It is suspected that the output depends on a quadratic (second degree) relationship between these variables. Write down the feature vector. What is the dimension of the feature space?

Answer

Since the output is governed by a second-degree relationship, all the monomials of degree two are considered as features. The feature vector is written as $[x_1^2 , x_2^2 , x_3^2 , x_1 \times x_2 , x_1 \times x_3 , x_2 \times x_3]^T$. Here, the superscript denotes transpose of the matrix and is written in this form to save space by converting the column matrix into a row matrix. The feature vector consists of six elements. Hence the feature space is of dimension 6. The original input space is 3-dimensional.

Exercise 8.5 Deflection in a simply supported beam

Figure 8.2 shows a simply supported beam under a point load P. L is the span, E the Young's modulus of the material, and I the moment of inertia of the cross section. The deflection at the midspan of the beam depends on the E, I, P, and L. Suppose that the value of E is not known accurately. There is uncertainty in the values of other variables as well because of approximations in the geometrical model and boundary conditions. Uncertainties in material properties and boundary conditions are common in engineering. Experiments are conducted for different values of P, and the deflections are measured. Can the relationship between deflection and the other parameters be determined through linear regression?

HINT

Use non-linear features as input variables for regression.

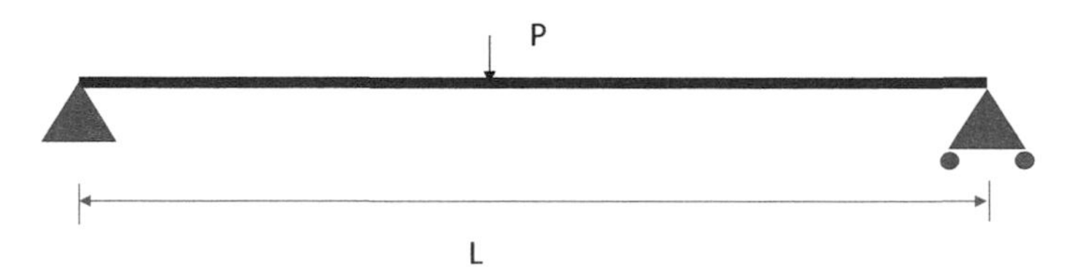

Figure 8.2 Simply supported beam under point load

8.5 Artificial neural networks (ANN)

ANN uses a complex mathematical model for regression. It can be viewed as a network of equations and is capable of representing highly non-linear and discontinuous relationships. Even though, it was initially developed based on analogy with biological nervous systems, today the analogy is not considered as important; the mathematical representation and the learning algorithms have progressed beyond mimicking the biological systems.

An ANN consists of a network of processing units called neurons (Figure 8.3). Each neuron has a set of incoming connections and a set of outgoing connections. (Figure 8.4). The neuron receives input from its incoming connections, and the output from the neuron is a function of the net input it receives. The output from a neuron is transmitted to other neurons through its outgoing connections, and the input received at the other end depends on the strength of the connection, known as the weight factor.

The neurons are typically arranged into layers (Figure 8.5). The input layer represents input variables. The output layer represents output variables. There might be one or more hidden layers containing nodes representing intermediate derived quantities. A two-layered network is shown in Figure 8.6. The input layer consists of two nodes representing the variables x and y. The output layer has a single node representing the output variable z. There are no hidden layers. w_1 and w_2 are the weights representing the strength of connections from the input nodes to the output node. The output from a node in the input layer is the value of that variable; this is provided by the user in the form of training data or test data. Similarly, the output from a node in the output layer is the value of the output variable predicted by the network.

A four layered network is shown in Figure 8.7. In this network there are four nodes in each of the hidden layers. The architecture of the network is defined by the number of layers, the number of nodes in each layer, and how they are connected; the architecture is defined by the user. The architecture determines what type of relationships can be represented by the network.

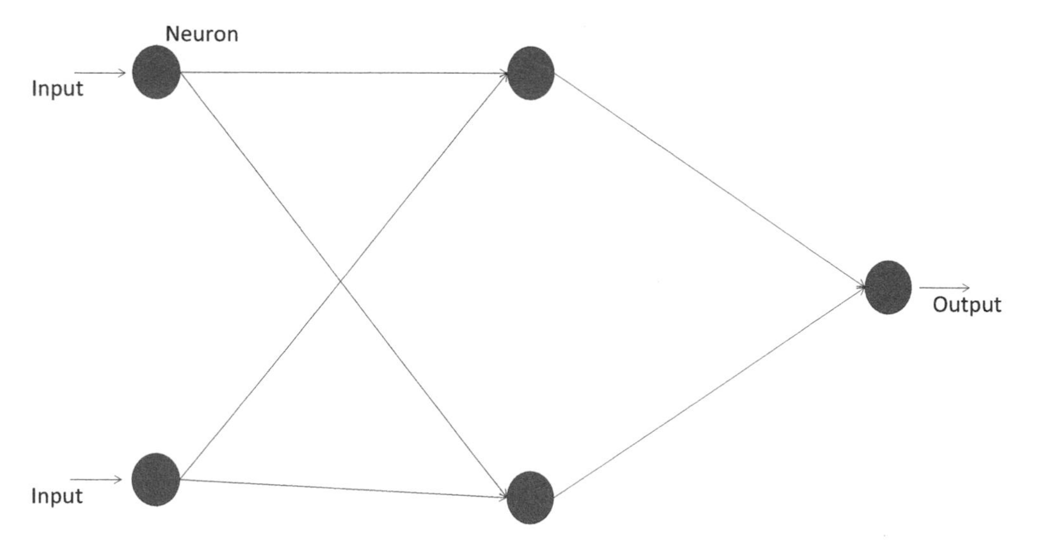

Figure 8.3 Artificial neural network

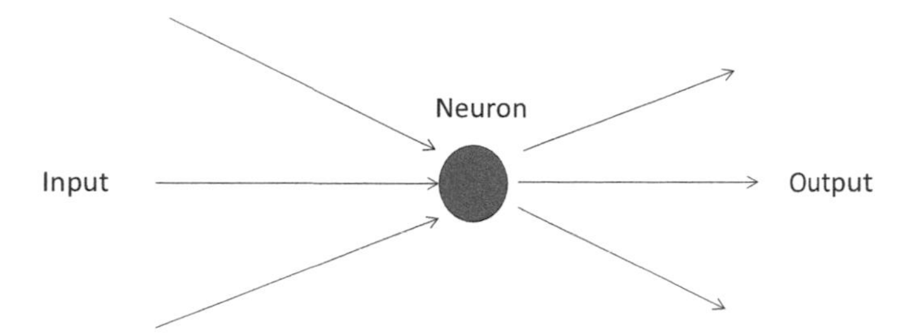

Figure 8.4 A neuron in an artificial neural network

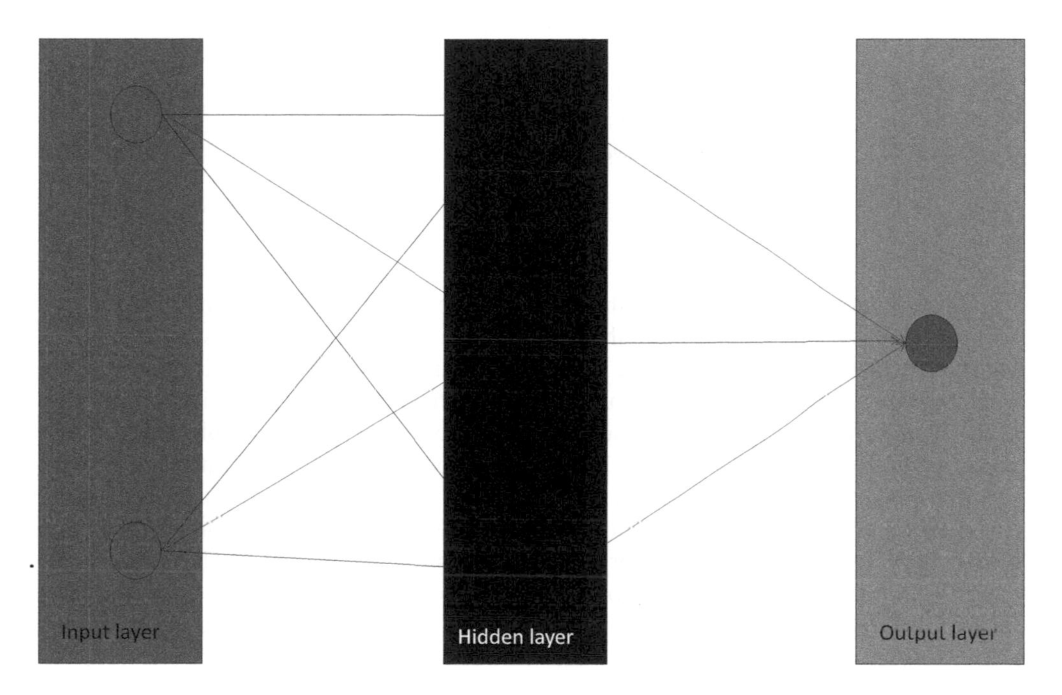

Figure 8.5 Layers in an artificial neural network

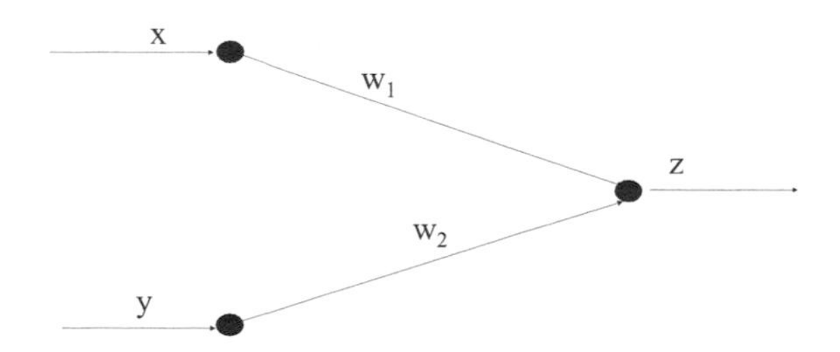

Figure 8.6 A two-layered network

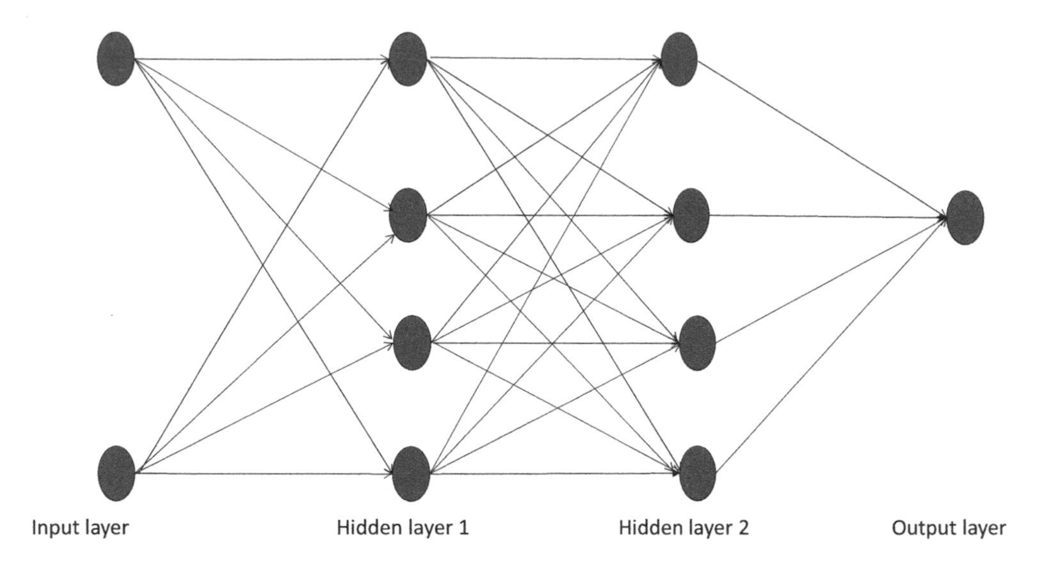

Figure 8.7 A four-layered network

The net input received at a node is the sum of the inputs from all the incoming connections (edges). This is the weighted sum of the output from all the connected nodes in the previous layer. In the network shown in Figure 8.6, the net input received by node z is (w_1 x + w_2 y). The output from the node depends on the activation function. A simple activation function is the identity function, that is, the output is equal to the input. A threshold function was proposed in the early days of ANNs. It produces an output only if the input is greater than the threshold. A two-layered network called as a *perceptron network* was used initially for image recognition tasks. A perceptron network outputs only two values, 1 or −1, depending on whether the net input is greater or less than the threshold value of 0. This is useful for classification problems where the output is discrete. However, the discontinuity in the output function causes difficulties for using optimization methods for minimizing the prediction error. Non-linear activation functions, which do not have this problem, are described here.

A sigmoid function is defined as follows:

$$\sigma(s) = \frac{1}{1+e^{-s}} \qquad \text{Eq 8.16}$$

This is also called a logistic function. If s is the net input at a node, the output from the node $\sigma(s)$ is given by the Eq 8.16. This function outputs values between 0 and 1. The function is plotted in Figure 8.8. For large values of s on the positive or negative side, the output is almost a constant, that is, zero or one. In this respect, the function is similar to a threshold function. However, this function is smooth and continuous, unlike the threshold function. Therefore, the function helps in using algorithms that make use of gradients for optimization.

Another popular activation function is ***ReLU*** standing for Rectified Linear Unit. It outputs 0 if the input is negative; otherwise, the output is equal to input. This is shown in Figure 8.9. Like a threshold function, it produces no output until the input is equal to a threshold (0). But unlike the threshold function, there is no discontinuity in the output.

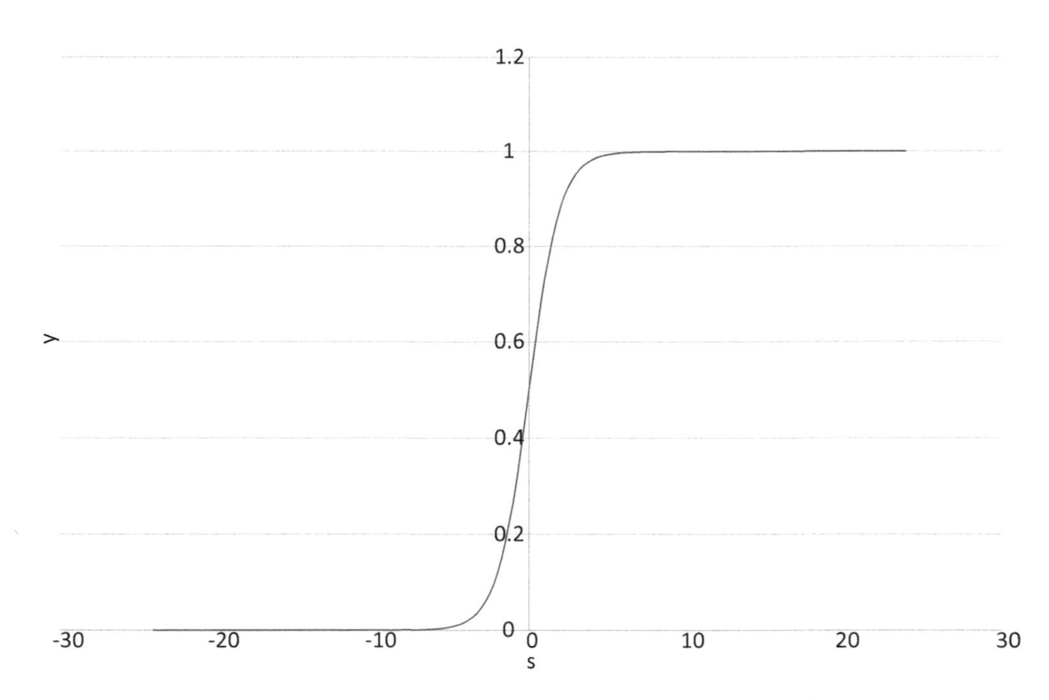

Figure 8.8 Sigmoid function

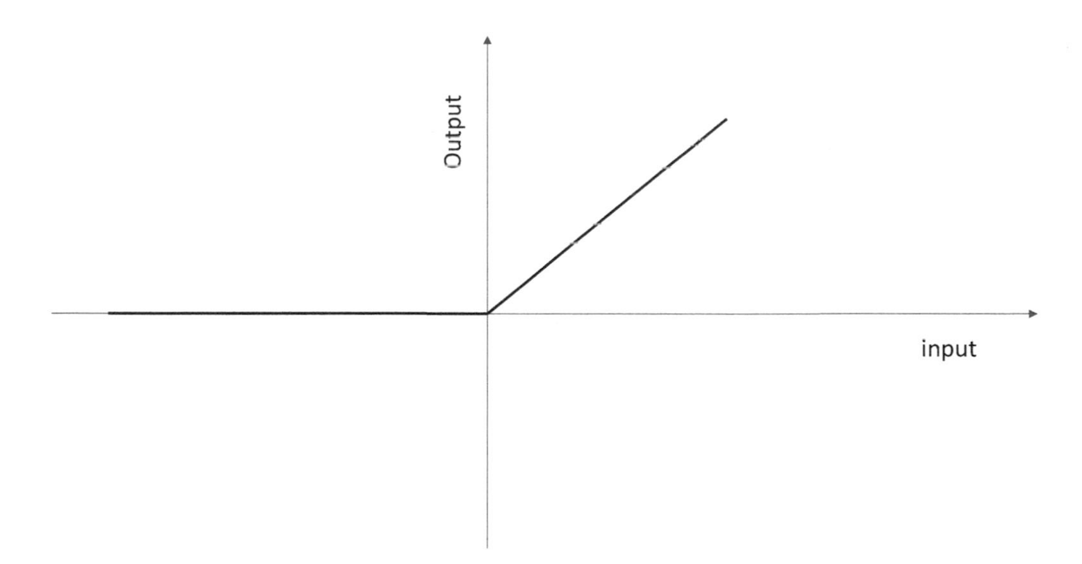

Figure 8.9 RELU function

Exercise 8.6

Determine the form of the equation represented by the ANN shown in Figure 8.10. The hidden layer uses the sigmoid activation function. The other layers use identity output. Comment about the nature of the relationship represented by this network. Can it represent a linear relationship?

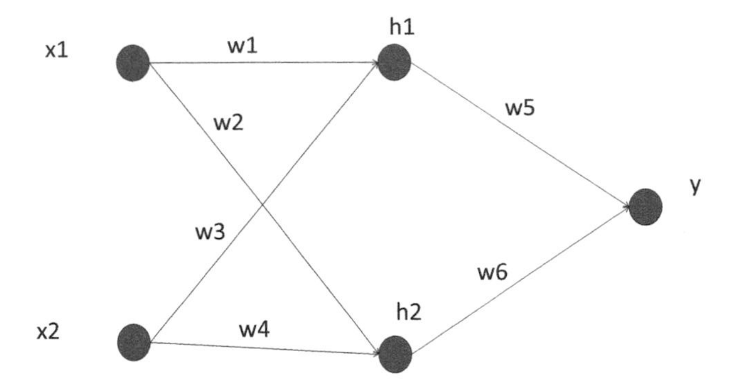

Figure 8.10 Determine the form of the equation represented by the ANN

Solution

The net input at the hidden node h1 is: w1 x1 + w3 x2. Since this node has sigmoid activation function, the output from this node is:

$$\frac{1}{1+e^{-(w1x1+w3x2)}}$$

Similarly, the output from the node h2 is:

$$\frac{1}{1+e^{-(w2x1+w4x2)}}$$

The output y is the weighted sum of the output from the hidden nodes. Therefore

$$y = \frac{w5}{1+e^{-(w1x1+w3x2)}} + \frac{w6}{1+e^{-(w2x1+w4x2)}}$$

In general, this is a non-linear relationship between y and the input variables (x1, x2). However, when the weight factors (w1, w2, w3, w4) are small, the expression can be written as a Taylor's series expansion involving linear and non-linear terms. By fine tuning the values of the weight factors, the expression can approximate linear relationships.

8.5.1 Learning algorithms for ANN

Learning in ANN occurs by updating the weights of connections. Training data are used to update the weights such that the difference between the value of the output variable in the training data and the output produced by the ANN is a minimum. In general, this is an optimization problem in which a loss function is minimized. There are many learning algorithms such as Hebb's rule, Hopfield's law, and Widrow-Hoff rule. These algorithms differ in the heuristic that is used to minimize the difference between the output of the ANN and the target output. In multi-layered networks, a form of back-propagation algorithm is used to update the weights. Typically, gradient descent is used to minimize the loss function. It works by computing the error in the predictions made by the network, that is, the error in the output layer. Then the errors are propagated to the previous layers. Finally, the weights are updated proportional to the error. Since this procedure involves gradient descent, it is not guaranteed to find the global minimum of the optimization problem, it might converge to a local minimum. However, for many practical problems, excellent results are obtained, and that is the reason for the popularity of this form of learning.

8.5.2 The hypothesis space in ANN – its representational power

The mathematical model that is used to make the prediction in ANN is very complex (as illustrated in Exercise 8.6). The types of mathematical equations that can be represented by the model depends on the network architecture as well as the model parameters such as the weights of edges, the bias value of the nodes, and the activation function used. The simplest architecture is a two-layered perceptron network shown in Figure 8.6. There are no hidden layers in this network; there is only one input layer and one output layer. The output layer uses a threshold function; it outputs only +1 or −1 depending on whether the input is greater than or less than 0. Such a network can only represent certain types of relationships between the input and output variables. First of all, the output value is discrete because of the use of a threshold function. Hence it is used mainly for classification problems such as image recognition. Secondly, it is capable of representing certain types of Boolean relationships such as AND, OR, NOT AND, and, NOT OR. The AND operator outputs 1 only if all the input variables have value 1, otherwise, it outputs 0. The OR operator outputs 1 if at least one of the input variables has value 1. NOT is the negation of a condition, that is, it returns the opposite of the input value. These are common conditions that could affect the outcome of an engineering decision-making. For example, we operate a tower crane only when it is not windy AND there is no heavy rain. Therefore, if an ANN has to learn to produce this output correctly, the network should be capable of representing this relationship. Of course, the threshold function should be modified to output 0 instead of −1, if the output expected from the Boolean operators are to be produced. It can easily be verified that by playing with the weight factors in the network shown in Figure 8.6, the required output values for AND OR conditions can be obtained for any combination of values for input variables.

Even though the two-layered perceptron can represent AND OR conditions, it cannot represent an *exclusive or* relationship (XOR). When the XOR operator is applied to two input variables, it outputs 1 if only one of the variables has value 1 and it outputs 0 if both variables have value 1 or 0. Practical applications of this operator seem rare. However, it is a fundamental operator in computer systems and important in correctly modeling complex operations. Isolated practical engineering examples like the following can also be found: If two robots are permitted to pick up an object at the same time, both will obstruct each other, and the output will be nothing. If only one of them is permitted, the work gets done. If there are no robots available, again, there will be no work done. These conditions are represented by the XOR operator.

The two-layered perceptron network is incapable of representing the XOR condition. However, a multi-layered network can do this. An example is shown in Figure 8.11. Thew-weights of all the edges are shown in the figure. The hidden and output layers use threshold function with a threshold value of 0.5.

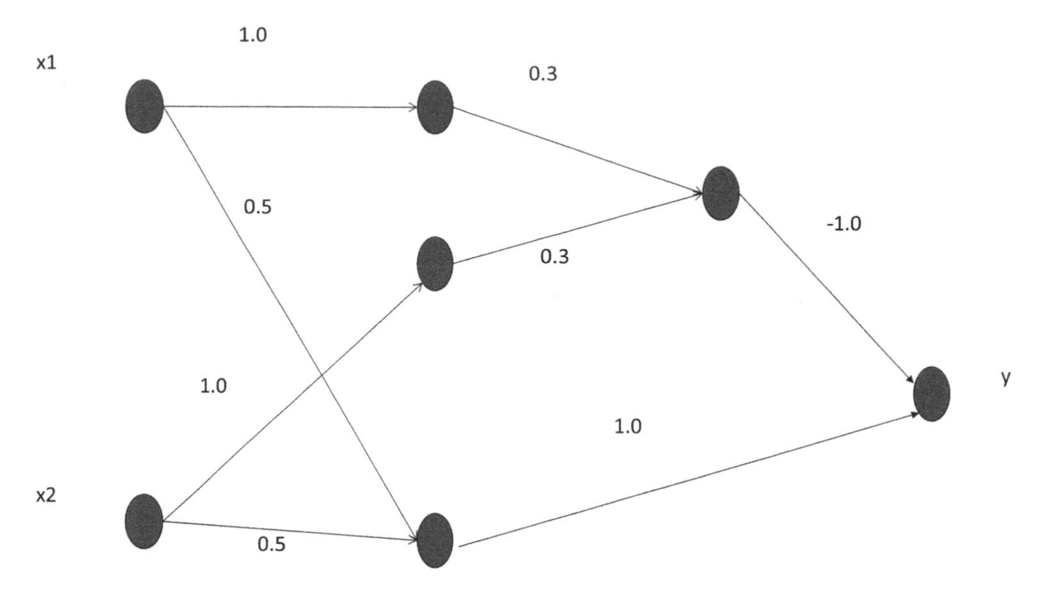

Figure 8.11 A multi-layered network that represents XOR relationship

Exercise 8.7

In Figure 8.11, Calculate the output from node y for the following combinations of values for x1 and x2. What is the relationship represented by the network?

x1	*x2*
0	0
0	1
1	0
1	1

Since discontinuous output can be produced using threshold functions, the ANN can represent discrete output variables. Hence, it could be used for classification problems as well as classical regression problems. Figure 8.12 Is an example of a network that outputs a discrete variable, the slab system for a room. There are three possible output values, one-way, two-way, and waffle slabs. The input variables are the span and the shape of the room. These are converted into binary variables by choosing discrete values: large span, small span, square shape, and rectangular shape. The output node that has the maximum value is taken as the recommended slab system for a given combination of values for span and shape.

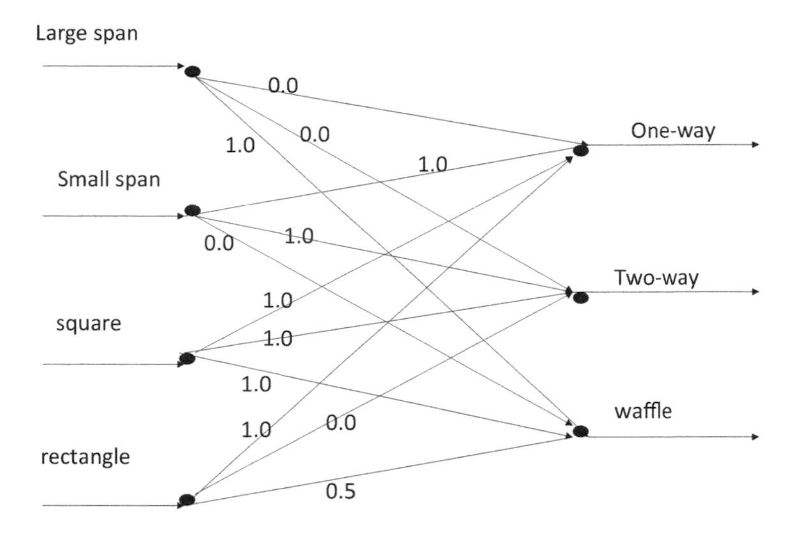

Figure 8.12 A two-layered network for a classification problem

8.6 Summary

Regression is used where the exact analytical model for predicting an output variable is not available or reliable. In such situations, an empirical model can be developed through regression using measured or simulated data. Regression is also used to create surrogate models when simulations are time-consuming. Ordinary least squares linear regression is the simplest statistical learning technique in which the coefficients of input variables are obtained by solving a system of linear simultaneous equations. However, it does not work well when there are linear relationships between input variables in the data (multi-collinearity) or when there are more variables than the number of data points. Other methods such as ridge regression might be applicable in some cases. Artificial Neural Networks (ANN) are a means of performing non-linear regression. It can also be used for classification tasks by treating the task as a regression problem involving discrete output variables. A loss function, typically using mean square error, is used to update the weights in the network such that the predicted output matches the output in the training data.

References

Stephen, Stefie J., Raphael, Benny, Gettu, Ravindra, and Jose, Sujatha. (2019). Determination of the Tensile Constitutive Relations of Fiber Reinforced Concrete Using Inverse Analysis, *Construction & Building Materials*, 195, pp. 405–414. https://doi.org/10.1016/j.conbuildmat.2018.11.014

Ward, L.G. and Shakespeare, R. (1998). *Rendering with Radiance: The Art and Science of Lighting Visualisation*, San Francisco, CA: Morgan Kaufmann.

9 Classification task

9.1 Introduction

The classification task was qualitatively introduced in Chapter 6. This chapter discusses more details and presents selected algorithms for this task. Since a deeper understanding of the algorithms is not possible without mathematical details, essential equations are included in this chapter. Readers who are not comfortable with mathematics might skip the mathematics and glance through the important observations presented as boxed text. Traditional binary linear classification, maximal margin classification, support vector machines, and non-linear classification using kernel methods are covered in this chapter. Decision trees and random forests are other examples of algorithms for classification. These are discussed in the Chapter 10.

9.2 Image recognition

An example of image recognition is discussed first to understand the classification task better. Image recognition has several applications in building and construction. Automated navigation on construction sites needs information about objects that are in the path of the vehicle or robot. Precise alignment of parts for fabrication and assembly requires detection of features such as edges and holes. Access control to buildings might make use of face recognition. Counting occupancy in a room could also be done taking pictures and recognizing faces. There are many other examples of the use of computer vision in building and construction automation.

A simple example of recognizing a rectangle in an image is taken first to illustrate key concepts. In Figure 9.1, the pixels of an image containing a rectangle are shown. There are 32 pixels horizontally as well as vertically. The pixels forming the rectangle are black, and all other pixels are white. Data for one image consists of the values of 32×32 pixels, that is, 1,024 values. Each pixel is given a value of 1 if it is white and 0 if it is black. Given this data, we want to determine whether a rectangle is present in the image.

The task of recognizing a rectangle may be performed as follows.

1. Prepare training data containing images of rectangles and other shapes. Each image is labeled positive if it contains a rectangle. It is labeled negative otherwise.
2. Develop a machine-learning model for a binary classification task. The input to this model consists of values of 1,024 pixels. That is, there are 1,024 input variables, each taking a value, either one or zero. The output is either 1 or -1 depending on whether the image contains a rectangle or not.

DOI: 10.1201/9781003165620-11

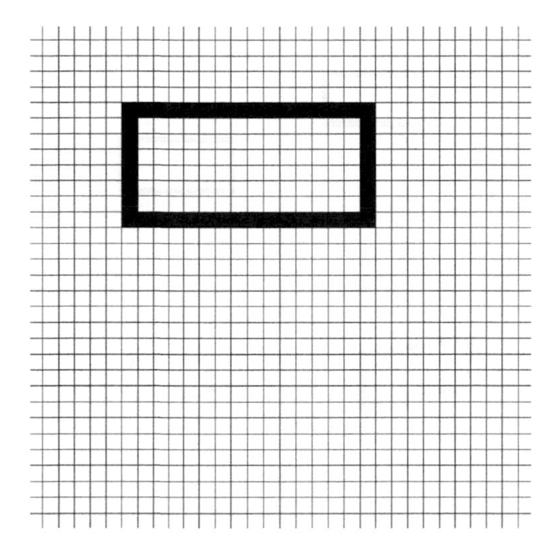

Figure 9.1 Pixels of an image containing a rectangle

3 Train the machine learning model using the data prepared in Step 1.
4 After finishing the training, present a new image to the model. The model outputs 1 or −1, indicating the presence or absence of a rectangle

In realistic cases, several variations to the image are possible. The rectangle could be transformed by translation, rotation, or scaling. In addition, the thickness of the rectangle could be one or more pixels. A few variations are shown in Figure 9.2. The pixel values change completely after the transformations. Still, the machine learning system should be able to recognize these as rectangles. In fact, many combinations of pixels are possible through various transformations. It is not possible to input all these images into the training set. With a few input images, the algorithm should be able to generalize and learn to recognize rectangles. It is clear that raw pixel values cannot be compared to recognize rectangles effectively; the pixel values will be different for each image. Learning should involve high-level generic features, rather than raw pixel values. The patterns that are observed should be generalized such that unseen variations of the images are also classified correctly.

Exercise 9.1

1 How many input variables are there if the image is not pure black and white, but has different shades of gray?
2 How many input variables are there if the image is in color, having different values for red, green, and blue?
3 How many input variables are there if the size of the image is 16 × 16, that is, 16 rows of pixels in each direction?
4 Will the learning be affected, if only positive examples are input for training? That is, only images of rectangles are provided.

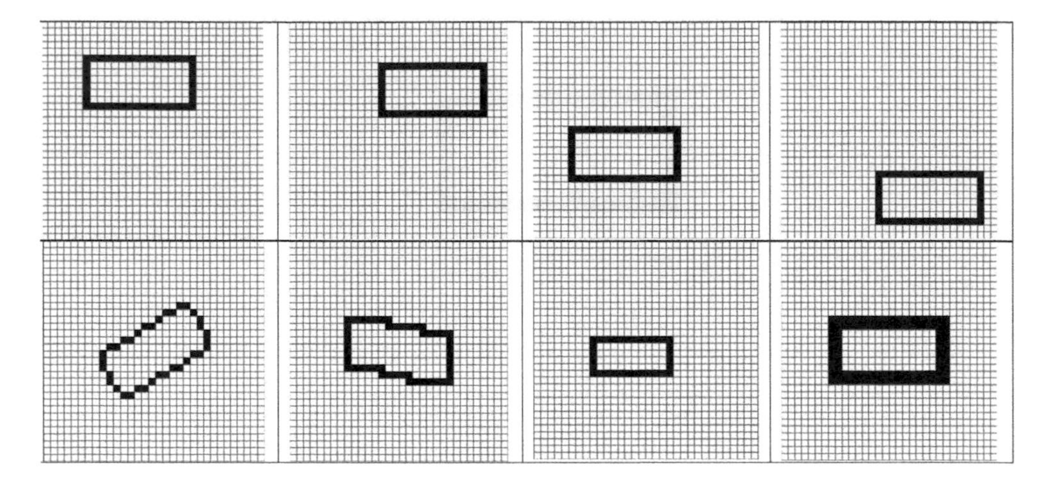

Figure 9.2 Pixels of a rectangle after various transformations

9.3 Binary linear classification

A binary linear classifier groups data points into two classes, positive and negative. These points are separated by a hyperplane whose equation is given by

$$\Sigma w_j x_j + b = 0 \qquad \textit{Eq 9.1}$$

Here, x_j is the j-th variable and w_j is the coefficient of the j-th variable. b is called the bias term, which produces an output even when all the input values are zero. This equation is written using vector notation as:

$$\langle \boldsymbol{w}, \boldsymbol{x} \rangle + b = 0 \qquad \textit{Eq 9.2}$$

Here, **w** and **x** are vectors of dimension *n*, *n* is the number of input variables. Note that vectors are written in bold face. The vector **x** contains all the input variables x_j, and **w** contains the corresponding weight factors w_j. The angle bracket represents the dot product of these two vectors. This is computed by multiplying each weight factor w_j by the corresponding value of x_j and summing up all the products. This is a convenient short form of Eq 9.1.

The coefficients w_j are determined such that all the positive points in the training data satisfy

$$\langle \boldsymbol{w}, \boldsymbol{x} \rangle + b > 0 \qquad \textit{Eq 9.3}$$

All the negative points satisfy

$$\langle \boldsymbol{w}, \boldsymbol{x} \rangle + b < 0 \qquad \textit{Eq 9.4}$$

These two conditions can be combined into a single expression as follows:

$$(y_i(< \boldsymbol{w}, \boldsymbol{x}_i > +b)) > 0 \qquad Eq\ 9.5$$

Where, the subscript i represents the index of a data point (it is not the index of an input variable). The vector $\boldsymbol{x}_i$ contains the values of all the input variables of the i-th data point. y_i is the output value for the i-th data point $\boldsymbol{x}_i$. Note that y_i is a scalar quantity taking the value 1 if the point belongs to the positive class, and −1 for the negative class. For a positive class, both the factors that are multiplied on the left-hand side of Eq 9.5 are positive. So, the product is greater than zero. For a negative class, both the factors are negative; again, the product is positive. Thus, the equation represents both the conditions Eq 9.3 and Eq 9.4.

Exercise 9.2 Equipment breakdown classification – 1 variable

Table 9.1 contains past data about a type of automation equipment. The first column contains the age of the equipment, and the second column indicates whether it has suffered a breakdown in the past. Pose this as a binary linear classification problem. By trial and error, find the weight factor and bias such that Eq 9.5 is satisfied.

Answer

This problem involves one input variable (age) and one output variable (breakdown). Calling these variables x1 and y, the table is converted into the form in Table 9.2 after setting the value of y to 1 if there is a breakdown and −1 otherwise.

Positive points should satisfy

$$w1 \times x1 + b \geq 0$$

Negative points should satisfy

$$w1 \times x1 + b < 0$$

Table 9.1 Equipment Breakdown Classification

Age	*Breakdown?*
4	No
1	No
6	Yes
8	Yes

Table 9.2 Equipment Breakdown Classification Data

x1	*Y*
4	−1
1	−1
6	1
8	1

There are many possible values for the weight factor w1 that satisfy these equations. The solution is not unique. It can easily be verified that the assignment of w1 = 1, b = −5 satisfies all the equations. Thus, we write Eq 9.5 as:

$$\left(y_i\left(x_i - 5\right)\right) > 0$$

This inequality should be satisfied for each row in Table 9.2, where y_i denotes the y value for each row, and x_i denotes the corresponding x1 value.

Exercise 9.3 Equipment breakdown classification – 2 variables

This is a modified version of Exercise 9.2. Table 9.3 contains past data about a type of automation equipment. The first column contains the age of the equipment, the second column whether it is fully automatic, and the third column indicates whether it has suffered breakdown in the past. Pose this as a binary linear classification problem. By trial and error, find the weight factor and bias such that Eq 9.5 is satisfied.

Answer

There are two input variables in this problem, let us call them x1 and x2. x2 is converted into numbers 0 and 1; 1 indicates a fully automatic machine and 0 otherwise. The data are now presented using numeric variables in Table 9.4.

Comparing the data with the previous exercise, it is noted that adding the new variable x2 does not make any difference to the classification. Hence the same solution is valid. However, since there are two variables, the weights are written in vector notation as follows:

$$w = \begin{bmatrix} 1 \\ 0 \end{bmatrix}$$

Eq 9.5 is written as:

$$\left(y_i * \left(< \mathbf{w}, \mathbf{x}_i > - 5\right)\right) > 0$$

Table 9.3 Equipment Breakdown Classification – Two Variables

Age	*Fully automatic?*	*Break-down?*
4	Yes	No
1	No	No
6	No	Yes
8	Yes	Yes

Table 9.4 Data for Equipment Breakdown Classification – Two Variables

x1	*x2*	*y*
4	1	−1
1	0	−1
6	0	1
8	1	1

This should be satisfied for each row in Table 9.4, where y_i denotes the y value for each row and $\mathbf{x}_i$ denotes the corresponding input vector consisting of the values of x1 and x2. For verification, the calculations are shown in Table 9.5.

In this example, the weight factor for x2 is 0. This means that this variable is not useful in determining the output. That is, whether the equipment is fully automatic or not does not determine whether it is likely to break down. By examining the relative values of weight factors, we can determine how useful the variable is in predicting the output.

9.3.1 Geometric interpretation

Eq 9.2 represents a hyperplane in *N* dimensions, the equivalent of a planar surface in three dimensions. The function that separates the points into different classes is known as the ***decision boundary***. In *N* dimensions, it is a hyperplane for linear classification tasks. In two dimensions, it is a straight line that separates points into positive and negative classes. See Figure 9.3. The input variables are represented by the horizontal and vertical axes x_1 and x_2. The class label is denoted by the symbols + and −. The thick dark line separating the positive and negative points has the equation of the form

$$w_1x_1 + w_2x_2 + b = 0$$

which is written in vector notation as Eq 9.1.

The vector **w** defines a direction perpendicular to the hyperplane. The parameter b determines how much the hyperplane is shifted from the origin. Varying the value of b makes

Table 9.5 Data for Equipment Breakdown Classification – Two Variables

x1	*x2*	*y*	*<w,x>*	*<w,x> + b*	*y × (<w,x> + b)*
4	1	−1	4	−1	1
1	0	−1	1	−4	4
6	0	1	6	1	1
8	1	1	8	3	3

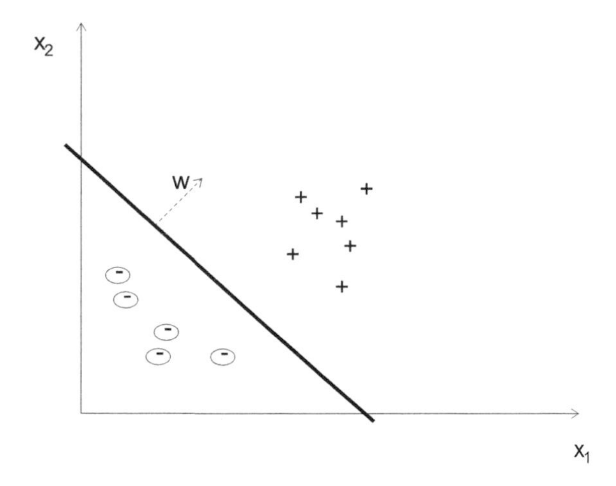

Figure 9.3 Binary linear classification. The weight vector **w** denotes the direction perpendicular to the hyperplane separating positive and negative points.

the hyperplane move parallel to itself. The absolute value of the function $(\langle \boldsymbol{w}, \boldsymbol{x}_i \rangle + b)$ is proportional to the distance of a given point *i* from the hyperplane. The minimum absolute value of this function determines how well the points on either side of the hyperplane are separated. It is a good indicator of the generalizability of the classifier. If points are too close to the hyperplane, there is high probability that a new point with a slightly different set of values of the input variables might end up on the wrong side of the hyperplane, resulting in misclassification.

Since the decision surface separating positive and negative classes is linear, it is called a ***linear discriminant***. Neural network researchers called it a perceptron.

9.3.2 Rosenblatt's Perceptron

Learning involves determining the values of weights and bias. Several algorithms have been developed for computing these values through iterative modifications using training data, one point at a time. With each data point, the algorithm checks the error in classification prediction and adjusts the weight vector. A classical algorithm for doing this is the Rosenblatt's Perceptron. It is shown in the following pseudo-code:

Loop 1: Repeat until there are no misclassifications

Loop 2: Repeat for each data point i

If $(y_i (\langle \mathbf{w}, \mathbf{x}_i \rangle + b) \leq 0)$ then

w is incremented by $\eta * y_i * x_i$
b is incremented by $\eta * y_i * R^2$, where R^2 is the maximum of x_i^2.

End of Loop 2

End of Loop 1

The inner loop in the algorithm (Loop 2) tries to reduce the classification error for each data point. It checks whether Eq 9.5 is satisfied. If it is not satisfied, it means that the current values of weight factors are not good, resulting in the misclassification of the current point. The weights must be updated. A heuristic is used to update the weight vector, which is given by this equation:

$$\Delta \boldsymbol{w} = \eta * y_i * \mathbf{x}_i \qquad \textit{Eq 9.6}$$

This is a vector equation; each element of the vector **w** is updated using the corresponding value of the variable in $\mathbf{x}_i$. That is, the increment to each weight factor is proportional to the corresponding value of the input variable in the vector $\mathbf{x}_i$ of the misclassified point. The variables that have low values in the current point are not significantly responsible for the misclassification since their product with the weight factor does not contribute much to the final output. Hence, their weights are not incremented much.

The product on the right-hand side contains y_i, which is either 1 or −1. This factor chooses the direction in which the weight is adjusted. If a positive point is misclassified, a positive increment is given to the weights, (if the x values are also positive). This ensures that the dot product <w,xi> increases, eventually causing the point to be classified as positive. Similarly, if a negative point is misclassified, a negative increment is given to the weights. Hence, the heuristic aims to reduce the number of misclassifications, one point at a time. The bias term is updated by a slightly different expression. Since this is a constant, it is not multiplied by the value of any input variable; instead it is multiplied by a constant factor x_l, which is the square of the maximum length of the input data points.

The first variable on the right-hand side, η, is called the *learning rate*. It is a small number that controls how fast the weights are changed in each iteration. It affects convergence and stability of the process. A large learning rate causes rapid changes to the weights, and the solution may not converge. Weights could oscillate in each iteration and the prediction performance may not improve. Too small of a learning rate would require many iterations to remove all the misclassifications.

This heuristic learning algorithm is equivalent to rotating the hyperplane by small amounts such that misclassified points are eventually moved to the correct side of the hyperplane. Updating the bias term is equivalent to translating the hyperplane.

The algorithm converges if the data are linearly separable, that is, if there exists a hyperplane that separates the positive and negative points. See Cristianini and Taylor (2000) for the proof. If the points are mixed on both sides, and there is no possibility of finding a hyperplane that separates all the points, the algorithm ends up with oscillations in the values of the weight factors without converging. Even in the case of linearly separable data, the problem is not "well-posed". There are many possible decision boundaries producing the correct classification. These are hyperplanes that are rotated slightly such that they still do not produce any misclassification. See Figure 9.4. In this figure, each of the thick dark lines represent a hyperplane that correctly classifies all the data points. The perceptron algorithm stops at finding one hyperplane (out of the many possible) that correctly classifies positive and negative points. There is no guarantee that this classifier will correctly classify future points (generalizability is poor).

One important advantage of the perceptron algorithm is that it is memory efficient. The procedure does not involve manipulating large matrices. If memory is not sufficient, each point might be read one at a time from a secondary storage device to compute the prediction and to update the weights. The point could be discarded after that, and the process is repeated. This procedure is called as *online learning*, in contrast to batch learning in which all the data points are presented as one batch. Even though modern computers have much

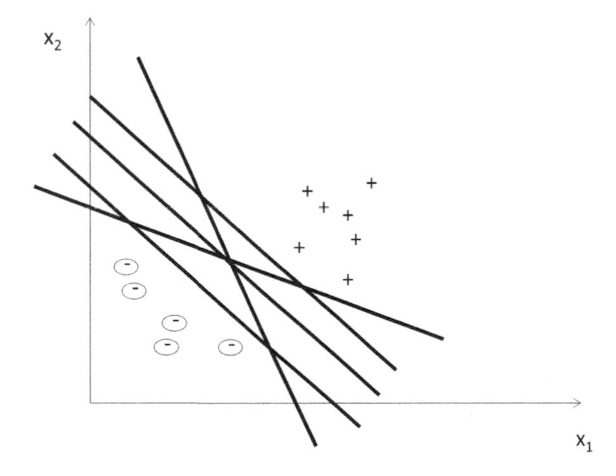

Figure 9.4 Multiple hyperplanes that correctly classify all the possible data points

larger memory, much larger problems are being solved today, creating higher demands on memory and making memory efficiency still important.

Exercise 9.4

Consider the data in Exercise 9.3. Calculate the weight factors using Rosenblatt's algorithm.

Answer

This numerical exercise is better done using a spreadsheet. Let us choose the learning rate to be 0.1. Let the initial weights and bias be equal to 0. First, we compute the R value, which is the maximum value of the norm of the input vectors. The R value is 8.12.

With the initial weights, the predicted value of the output variable is computed for each data point. If the point is misclassified, the weights are updated according to the algorithm. For the first point, (<w,x> + b) gives 0. Therefore, the weights are updated. The calculations are shown in Table 9.6. Within the fourth iteration, all the predicted values have the same sign as the output class.

9.3.3 Gradient descent

The algorithm presented in 9.3.2 can be rewritten as a gradient descent procedure to minimize an appropriate error function. In general, the error in a data point is the difference between the actual value and that predicted by the model. That is,

$$e_i = (y_i - y_{p,i}) \quad \textit{Eq 9.7}$$

where e_i is the error in the i-th data point, y_i is the actual output value in the data set, and $y_{p,I}$ is the value predicted by the model using the values of the input variables in the vector $\mathbf{x}_i$. In the case of the linear classifier, the predicted value is $\langle \mathbf{w},\mathbf{x}_i \rangle$. Note that y_i takes only two possible values, 1, or −1, depending on whether the class label of the point is positive or negative. The predicted value can be any positive or negative number and depends on the current weight factors in the model.

It is common to minimize the sum of the squares of errors, SE. This is also known as the square-loss function. This is computed by calculating the error in each data point, taking its square, and summing it up for all the points, as follows:

$$SE = \Sigma_{i=1}^{m}\left(y_i - y_{p,i}\right)^2 = \Sigma_{i=1}^{m}\left(e_i\right)^2 \quad \textit{Eq 9.8}$$

Table 9.6 Illustration of Perceptron Algorithm

	Iteration 1				*Iteration 2*				*Iteration 4*			
Point	*1*	*2*	*3*	*4*	*1*	*2*	*3*	*4*	*1*	*2*	*3*	*4*
<w,x> + b	0	−7	−9	−5.1	−2.6	−5.6	−0.6	12.8	−1.9	−5.4	0.6	2.9
w1	−0.4	−0.4	0.2	1	1	1	1.6	1.6	1.2	1.2	1.2	1.2
w2	−0.1	−0.1	−0.1	0	0	0	0	0	−0.1	−0.1	−0.1	−0.1
B	−6.6	−6.6	−6.6	−6.6	−6.6	−6.6	0	0	−6.6	−6.6	−6.6	−6.6

This is a good candidate to be taken as the objective function for numerical optimization. Gradient descent is a fast method for optimization. The gradient is the direction in which the function has the maximum rate of change. To minimize a function, it is best to take a step in the direction of the negative gradient. The gradient is a vector in which each element is the partial derivative of the function with respect to the corresponding variable. If O is the objective function, the gradient vector is written as

$$\nabla O = \begin{bmatrix} \frac{\partial O}{\partial w_1} \\ \frac{\partial O}{\partial w_2} \\ \cdots \\ \frac{\partial O}{\partial w_n} \end{bmatrix} \qquad \text{Eq 9.9}$$

Using the equation for O, the partial derivative with respect to the j-th variable is written as:

$$\frac{\partial O}{\partial w_j} = \sum_{i=1}^{m} 2e_i \frac{\partial e_i}{\partial w_j} = \sum_{i=1}^{m} 2e_i \frac{\partial \left(y_i - y_{p,i}\right)}{\partial w_j} = -2\sum_{i=1}^{m} e_i x_{i,j} \qquad \text{Eq 9.10}$$

where $x_{i,j}$ is the value of the j-th variable in the i-th data point. The last part of Eq 9.10 comes from the following expression for the predicted value for the i-th data point:

$$y_{p,i} = \sum_{j=1}^{N} w_j x_j \qquad \text{Eq 9.11}$$

Evaluating the partial derivatives using Eq 9.10, the gradient vector is computed. Dividing each element of the vector by the norm of the vector, we get a unit vector in the direction of the maximum increase in the value of the objective function. To get a point (weight vector) at a distance L in this direction from the current point **w**, you need to add the vector $L\nabla O$ to the current point. Taking a step in the direction of the negative gradient results in incrementing the weight of the j-th variable as follows:

$$\Delta w_j = \eta \sum_{i=1}^{m} e_i x_{i,j} \qquad \text{Eq 9.12}$$

where η is a parameter that combines the values of the step size L, norm of the gradient vector and the constant value 2 in Eq 9.10. This is written in vector notation as follows:

$$\Delta \boldsymbol{w} = \eta \sum_{i=1}^{m} e_i \boldsymbol{x}_i \qquad \text{Eq 9.13}$$

If the weight vector is updated in multiple steps, that is, using the contribution from each data point i, the summation is removed, and the increment is written as

$$\Delta w = \eta * e_i * \mathbf{x}_i \qquad \textit{Eq 9.14}$$

Similarity of this equation with Eq 9.6 is striking. The increment to the weight factor is proportional to the error and the value of the input variable for the i-th point. In the perceptron algorithm, the equivalent of the error term is just ±1 for misclassified points. Due to the particular objective function adopted here, the increments to the weights are higher since the magnitude of error will be greater than 1. It results in steeper descent compared to the perceptron algorithm.

Similar to the perceptron algorithm, points are taken one by one to update the weights using Eq 9.14. The process is repeated until there are no misclassifications or convergence is achieved. This procedure is known as the Widrow-Hoff or Adeline algorithm.

Updating the weights and bias can be done in a unified manner by putting the variable *b* also in the weight vector as the last element. So the new weight vector would look like this: $\{w_1, w_2, \ldots, w_n, b\}$. In the input vector **x**, a constant value of 1 is added at the end such that the product of the two vectors gives the original equation ($<\mathbf{w},\mathbf{x}> + b$). That is the input vector now becomes $\{x_1, x_2, \ldots, x_n, 1\}$. With this modification, the equations derived earlier are valid for the bias term as well.

It is tempting to start the iterative procedure with initial weights of 1 or 0. These values need not give the best results. This procedure is found to work quite well:

Step 1: Compute the centroid of all the positive points, a+.

Step 2: Compute the centroid of all negative points, a−.

Step 3: Define the normal vector from a− to a+. This is obtained by subtracting a− from a+, and then dividing it by the norm of this vector. Set this as the initial weight vector. If the norm is zero or too small, it means that the positive and negative points are evenly distributed. In that case, it is better to restart the iterative procedure using a random weight vector.

Step 4: The initial bias term is computed using Eq 9.21. (This equation will be explained later.)

Step 5: Iteratively, update the weight factors using Eq 9.14 until there are no misclassifications.

9.3.4 Maximal margin classifier

The generalizability is poor in the perceptron learning algorithm because the algorithm terminates when it finds a hyperplane that does not cause any misclassification. As mentioned earlier, there are an infinite number of possible hyperplanes, and the chosen hyperplane need not be the one that has the best performance in terms of prediction accuracy on unseen data. This is because the problem is not well-posed. This drawback is eliminated using the maximal margin classifier. Before describing the concept, key definitions are given in the following:

Definition: The *functional margin* of a data point $\mathbf{x}_i$ with respect to a hyperplane (**w**,b) is defined as

$$\Upsilon_i = y_i * (< w, x_i > + b) \qquad \textit{Eq 9.15}$$

A functional margin greater than zero indicates correct classification. However, it does not indicate how far is a point from the hyperplane. The weight vector **w** and the bias term *b* can be multiplied by an arbitrary positive constant without changing the classification accuracy. Points which are positive will remain positive after scaling **w** and b together. Only if **w** is normalized (that is, when **w** is a unit vector), the dot product $\langle \boldsymbol{w}, \boldsymbol{x}_i \rangle$ will measure the distance in the direction perpendicular to the hyperplane.

Definition: *The geometric margin* measures the Euclidean distances of the points from the decision boundary in the input space. It is obtained from the same equation by normalizing (**w**,b) – by dividing by the norm of w.

Definition: *Margin of a hyperplane* with respect to a training set S is the minimum geometric margin among all the points in the set. This is the geometric margin of the point lying closest to the hyperplane.

Definition: The *margin of a training set S* is the maximum geometric margin over all possible hyperplanes. A hyperplane realizing this maximum is known as a maximal margin hyperplane. The size of its margin will be positive for a linearly separable training set.

If a hyperplane is rotated or translated, its distance to the data points will change and therefore, the minimum geometric margin will be different. The maximal margin hyperplane is the one which has the maximum margin among all the hyperplanes possible (Figure 9.5). The margin of this hyperplane *m* is taken as the margin of the training set.

Definition: *Canonical hyperplanes* are hyperplanes having functional margin equal to 1.

The key concept behind maximal margin classification is the intuitive observation that if the geometric margin is high, the points are well separated, and the learnt hyperplane is unlikely to produce misclassifications. This can be proved mathematically using optimization theory (see Cristianini and Taylor, 2000). Starting with Eq 9.1, we note that if we multiply the weight factors and bias by the same positive value, equations Eq 9.3 and Eq 9.4 are still satisfied by the positive and negative points respectively. This means that the weights are not uniquely defined, and we have freedom to choose the scale factor. So, we choose the weights and bias such that the functional margin is 1. That is, we consider only canonical hyperplanes. There will be at least one positive point and one negative point with functional margin 1. (Otherwise, we can move the hyperplane parallel to itself and increase the margin.) Let these points be denoted by the symbols x^+ and x^- respectively. They satisfy the following equations:

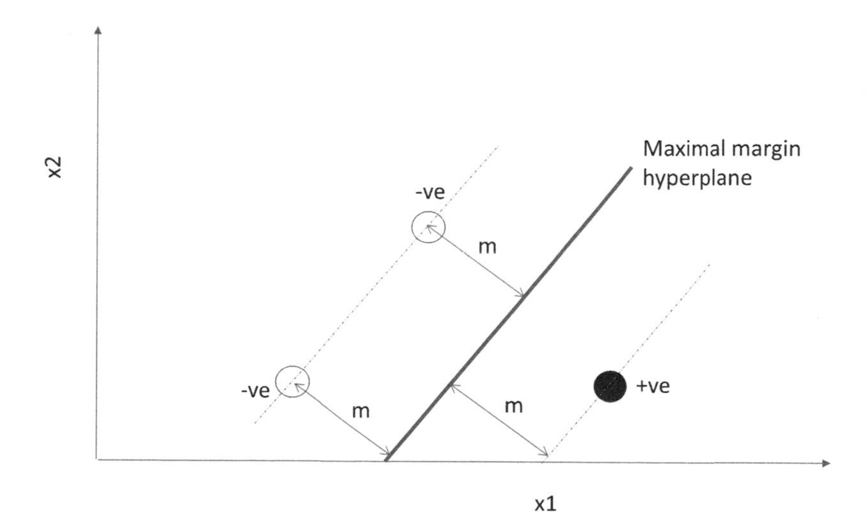

Figure 9.5 Maximal margin hyperplane

$$< w,x^+ > +b = +1$$
$$< w,x^- > +b = -1 \qquad \text{Eq 9.16}$$

Eliminating b by subtracting the second equation from the first.

$$< w,x^+ > - < w,x^- > = 2 \qquad \text{Eq 9.17}$$

The geometric margin is computed by defining a unit vector perpendicular to the plane and computing its dot product with the closest point. The unit vector is obtained by dividing the weight vector by the norm of the weight vector w. Since the dot product for the negative point is negative and has the same magnitude as the positive point, the geometric margin γ is taken as the average of the two as follows:

$$\gamma = (\frac{1}{2})\left(< \frac{\boldsymbol{w}}{\|w\|}, \boldsymbol{x}^+ > - < \frac{\boldsymbol{w}}{\|w\|}, \boldsymbol{x}^- > \right) = (\frac{1}{2})(\frac{1}{\|w\|})(< \boldsymbol{w}, \boldsymbol{x}^+ > - < \boldsymbol{w}, \boldsymbol{x}^- >) \qquad \text{Eq 9.18}$$

Substituting Eq 9.17 in Eq 9.18,

$$\gamma = \frac{\mathbf{1}}{\|w\|} \qquad \text{Eq 9.19}$$

Therefore, maximizing the geometric margin is equivalent to minimizing the norm of the weight vector for a hyperplane satisfying the conditions Eq 9.3 and Eq 9.4. Thus, the optimization problem can be written as

Minimize <**w**,**w**> subject to

$$(y_i * (< \boldsymbol{w}, \boldsymbol{x}_i > +b)) \geq 1 \textit{ for all data points } i \qquad \text{Eq 9.20}$$

Here, <**w**,**w**> gives the square of the norm. Minimizing this is equivalent to minimizing the norm. Note that the right-hand side of the constraint is 1 because we have assumed a functional margin of 1. Solving this optimization problem gives the hyperplane that maximally separates the positive and negative point.

The geometric margin is maximized by minimizing the norm of the weight vector for a hyperplane that does not cause any misclassifications.

While solving the perceptron problem, we had only the constraints shown in Eq 9.20. We did not have the objective function to minimize. Therefore, the solution was non-unique. On the other hand, solving the maximal margin classification problem yields a unique solution. Furthermore, the problem involves a quadratic objective function and linear constraints. This is a reasonably simple optimization problem to solve.

It is interesting to note that the maximal margin hyperplane minimizes the norm of the weight vector. This conforms Occam's principle described in Section 6.3.3. If we are able to reduce the weight factors of many variables to zero, these variables are not used for predicting the output. We get a model with minimum number of variables that explains the data. The simplest model is the best model according to Occam.

9.3.4.1 Solving the optimization problem

The maximal margin classification can be solved using the dual formulation presented in Section 9.6.1. A simple algorithm is presented here, which can easily be implemented in any programming language.

The optimization problem is solved in two stages. 1) constraint solving; 2) minimization of weights. The Adeline algorithm is used in Stage 1. A form of random search is used in Stage 2.

CONSTRAINT SOLVING ALGORITHM

The algorithm is presented in the form of the following pseudo-code:

Select learning rate η = 1, initial weights and bias = 0
Loop 1: Repeat until convergence
  Initialize the vector dw = 0, representing increments to weight factors
  Let number of misclassifications = 0
  Loop 2: Repeat for each data point
    Compute predicted output yp using the current weight factors. If there is any misclassification, add increment to dw using Eq 9.14, increment the number of misclassifications.
  If the number of misclassifications is 0, terminate Loop 1.
  If the number of misclassifications in this iteration is greater than or equal to that in the previous iteration, reduce the learning rate by multiplying by a factor, say, 0.5. If the learning rate becomes too small, convergence is achieved, terminate Loop1.
End of Loop1

At the end of this procedure, either a hyperplane is found that results in no misclassifications; or the learning rate becomes too small, making it impossible to improve any further. The second condition happens when the data set is not linearly separable. Stage 2 (weight minimization) is performed only if there are no misclassifications.

In the earlier procedure, the bias term should not be updated, it need not be taken as an optimization variable. It can instead by computed using the fact that the maximal margin plane is equidistant from the nearest positive and negative points. It follows directly from Eq 9.16. Thus, it is computed as:

$$b = -\left(\frac{1}{2}\right)[min(<\boldsymbol{w},\boldsymbol{x}^{+}>) + max(<\boldsymbol{w},\boldsymbol{x}^{-}>)] \qquad Eq\ 9.21$$

Where, x^+ denotes a positive point, and x^- denotes a negative point.

WEIGHT MINIMIZATION

Since the objective function is a simple quadratic function in the feasible region, an efficient random search method is used here. This is a random version of the method of bisection. To start with, the bounds for each variable (absolute value of weight factor) is set to be [0, w_c], where w_c is the weight factor that satisfies the constraints. The range [0, w_c] is the search interval in which the optimal value for w is sought. Since we are interested in minimizing the norm of the weight vector, the weight factors should be made as small as possible. Hence the minimum value is set to 0. The weight factors are updated, one variable at a time, by randomly generating a value in the range [0, w_c] and checking whether the constraints are satisfied. If the constraints are not satisfied, the generated value is discarded, and the bounds are updated. Otherwise, the new value is taken as the current optimal value and the process is continued. The process terminates when the search interval becomes too small, that is, when the minimum and maximum bounds are too close. The algorithm is written in the following pseudo-code:

Loop1: Repeat until convergence
Set the count the number of variables that have not converged to 0

Loop2: Repeat for each variable

If the search interval for the variable is too small, continue with the next iteration of Loop2

Increment the count the number of variables that have not converged by one

Generate a random value w_n for the weight of the current variable within the search interval. Weights of other variables are not changed. The bias term is recomputed using Eq 9.21. Compute the predictions for each data point.

If there are misclassifications, reset the weight of this variable to the previous value. Set minimum of the search interval to w_n. That is, the search interval is reduced so that it does not generate any more points below this value.

If there are no misclassifications, w_n is accepted as the current best value of weight for this variable. The maximum bound of the search interval is set to w_n, since higher values will increase the norm of the weight vector.

End of Loop2

If all the variables have converged, terminate Loop1

End of Loop1

9.3.5 Fisher's discriminant

Another method to make the classification problem well-posed is using the Fisher's discriminant *F*. It is defined as:

$$F = \frac{(m_1 - m_{-1})^2}{\sigma_1^2 + \sigma_{-1}^2}$$

where m and σ are the mean and standard deviation of the function output values (<w,x> + b) for the two classes indicated by the subscript. To explain the formula in a simpler manner, F is computed like this: calculate the mean of all the positive points m_1 and their standard deviation σ_1. Similarly, calculate the mean of all the negative points m_{-1} and their standard deviation σ_{-1}. Subtract the two means and divide by the sum of the standard deviations to get the value of *F*. *F* is maximized by finding the optimal values of **w** and *b*. This procedure will give the hyperplane that separates the means of the two classes by the maximum value. The optimization is performed using the standard method of setting the derivatives to zero. Even though, the procedure seems similar to maximal margin classification, the mathematical details are different.

9.4 Multi-class linear classification

There are many engineering problems where there are more than two classes to be separated. In the image recognition example, we have multiple objects to be identified. Then images of each object are assigned to a different class. In handwriting recognition, each letter is classified into one of the many classes representing different letters of the alphabet. In activity recognition, equipment and humans might be involved in several possible operations, and these have to be discerned.

When there are k classes to be separated, a simple technique is to convert it into k binary classification problems. In each problem, one class is taken as positive and all the remaining as negative. The hyperplane separating the positive examples from the rest are learnt as described in the previous section. This requires multiple training sessions, one for each class.

It is also possible to learn multiple classifiers in one training session. Binary linear classification principles can easily be extended to the case of multi-class as follows:

A set of weight factors, $\mathbf{w}_c$, and bias, b_c, is defined for each class c. That is, there is one hyperplane associated with each class. A point $\boldsymbol{x}_i$ is assigned to the class whose hyperplane is furthest from it. This is done by computing $(< \boldsymbol{w}_c, \boldsymbol{x}_i > +b_c)$ for each class and selecting the class which has the maximum value. Training can be done by gradient descent by minimizing the prediction error as described in the previous section.

9.5 Non-linear classification

In many applications, data are not linearly separable, that is, it is not possible to find a hyperplane that splits data into positive and negative classes. Non-linear decision boundary should be learnt. Machine learning techniques such as artificial neural networks can do this efficiently. Support vector machines using kernels are also proven to be efficient at this task. Before these are discussed, the fundamental concept of learning in the feature space is described.

9.5.1 Learning in the feature space

Section 8.4.1 introduced the concept of learning in the feature space. It was used for learning non-linear relationships in regression. The same concept can be used for classification. The derived features might be considered as new input variables, and the output variable can be written in the familiar form as

$$\sum_{j=1}^{p} w_j f_j(\boldsymbol{x}) + b = 0 \qquad \textit{Eq 9.22}$$

Where w_j is its coefficient of the j-th feature, $f_j(\mathbf{x})$. Note that the summation is over p elements, not n. By replacing the original input vector by the feature vector, the techniques we have discussed so far can be used to learn non-linear relationships. The decision boundary is linear in the feature space.

Exercise 9.5 Illuminance level in a room

A small skylight is at the center of a room. Brightness (illuminance) measurements were taken at random spots inside the room to determine whether there is excessive illuminance causing glare. A point is considered as negative if the illuminance level is too high. It is positive if the illuminance is within limits. In Table 9.7, coordinates of the points (x1, x2) in the room are given and their respective class labels. The origin is at the center of the room. Find out the decision boundary that separates positive and negative classes. The original input data are plotted in Figure 9.6. The decision boundary is a circle in this case. Now consider features (f1, f2), which are the quadratic terms ($x1^2$, $x2^2$). These are

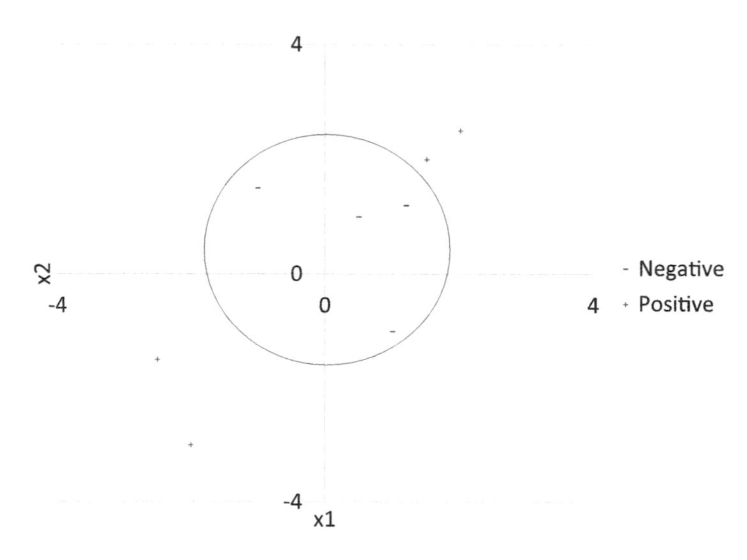

Figure 9.6 The decision boundary is non-linear in the input space for this data set.

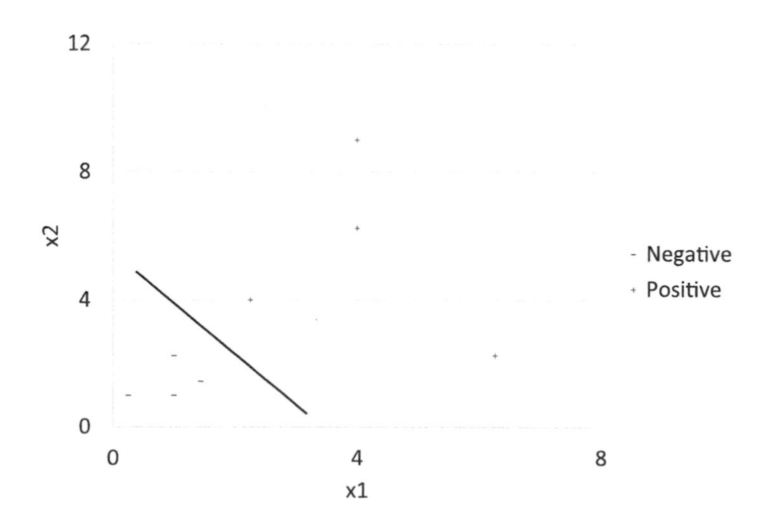

Figure 9.7 The decision boundary is linear in the feature space for the same data set in Figure 9.6.

Table 9.7 Coordinates of Points in a Room and Their Illuminance Level

X1	*X2*	*class*
0.5	1	−1
1.2	1.2	−1
−1	1.5	−1
1	−1	−1
1.5	2	1
2	2.5	1
−2.5	−1.5	1
−2	−3	1

plotted in Figure 9.7. Here, we can locate a linear decision boundary separating positive and negative classes.

Features must be selected using domain knowledge; features must be meaningful to the application. There are many possible ways of selecting the features. The best results are obtained by selecting a minimal set of features that could be explained using physical principles. Occam's principle is relevant here. While it is tempting to include many high-order monomials as features, it might result in overfitting and should be avoided, unless it is necessary for improving accuracy.

9.5.2 Image recognition

Image recognition is discussed in this section as an example of a classification task. Many difficulties in applying machine learning for practical problems are brought out through this example. Consider the image of the rectangles shown in Figure 9.1. As explained in Section 9.2, choosing the array of pixels as input variables is not appropriate for this problem. Every member of the positive class has completely different values for each pixel. Hence the problem becomes not linearly separable. Therefore, good features need to be extracted from the data. The features represent domain knowledge related to what are good characteristics of the positive set and what can potentially discriminate the negative set from the positive set.

9.5.2.1 Feature extraction

What are good features for image recognition? For recognizing rectangles, the presence of horizontal and vertical edges is important from our knowledge of geometry (this is the domain knowledge in this example).

Edge detection techniques are well developed in computer vision. Since image recognition is treated only as an application example this book, an exhaustive treatment of this is not attempted here. Simple concepts are presented for understanding the concept of feature extraction.

An edge is detected by the change in the value of the pixel, that is, the gradient of pixel value. When the pixel value changes from 0 to 1, that might indicate the location of a point on an edge. The gradient is computed by calculating the difference between adjacent pixel values, either in the horizontal direction or vertical direction. Better results are obtained by the central difference formula in which the next pixel value is subtracted from the previous value. If adjacent pixels are stored in a 3-dimensional array as,

$$\begin{bmatrix} p_{11} & p_{12} & p_{13} \\ p_{21} & p_{22} & p_{23} \\ p_{31} & p_{32} & p_{33} \end{bmatrix}$$

subtracting the third column from the first column indicates how much the next pixel has changed compared to the previous pixel. This can be written as a matrix operation as follows:

$$\begin{bmatrix} -1 & 0 & 1 \\ -1 & 0 & 1 \\ -1 & 0 & 1 \end{bmatrix} \begin{bmatrix} p_{11} & p_{12} & p_{13} \\ p_{21} & p_{22} & p_{23} \\ p_{31} & p_{32} & p_{33} \end{bmatrix} \quad \textit{Eq 9.23}$$

Here, the dot operator performs an operation similar to the vector dot product for one dimensional vector. For two-dimensional matrices, the inner dot product is obtained by multiplying corresponding elements in each row and taking the sum of all the products (note that this is not the familiar matrix multiplication rule in two dimensions). In Eq 9.23, this results in the expression, $((p_{13} - p_{11}) + (p_{23} - p_{21}) + (p_{33} - p_{31}))$, that is sum of differences in pixel values in each row. If all the pixels change in the horizontal direction, the maximum value of 3 is obtained. If no pixel changes across any row, 0 is obtained. Higher value indicates greater strength of the edge. The first matrix in Eq 9.23, is called the Prewitt vertical mask or filter, which is one of the many filters that have been developed for edge detection. To detect horizontal edges, the Prewitt horizontal mask is defined as:

$$\begin{bmatrix} -1 & -1 & -1 \\ 0 & 0 & 0 \\ 1 & 1 & 1 \end{bmatrix} \quad \textit{Eq 9.24}$$

To detect all the vertical edges in an image, the filter is applied to all the pixels in the image by repeating the operation for all the rows and columns, except the first one and the last one. The first and last rows must be omitted because the previous and next pixels are needed to apply the operator. At each pixel location, a number is obtained that denotes the strength of the edge at that point. This process is called the convolution operator. It is said that the mask is convolved over the entire image, that is, the same inner dot product is repeated from right to left and then top to bottom. Similarly, to detect all the horizontal edges, the horizontal mask is convolved over the entire image. Through this process, a new image is obtained in which only pixel locations where edges are present have non-zero values.

With this background in edge detection, we are ready to select good features for identifying rectangles in an image. Since the parts of the image where there are no edges are not important in detecting rectangles, we select only a small region which has the highest values for edge strength. Select a small window size, say 6×6 pixels. Within this window, we want to calculate a score that combines the strength of both horizontal and vertical edges, which potentially indicates the presence of a corner of the rectangle. A simple score function will be the sum of the edge strengths at each position within the window. The window containing the largest sum of weights is located by moving the window in the horizontal and vertical directions. The sum of strengths of all the edges in each window is calculated. The window that gives the largest sum is selected. The strengths of edges at each pixel location within this window are taken as the features for separating images into different classes. With a window size of 6 × 6, there are 36 input variables (features) and correspondingly, 36 weight factors. This is a huge reduction from 1,024 input variables in the original data. Furthermore, even images that are translated have similar values for the chosen features because the selected window will contain the rectangle; most likely a corner of the rectangle will be selected since that has the maximum strength of edges after summing up horizontal and vertical edge scores. With maximal margin classification, discriminating features of the input data are obtained by minimizing the weights of the input variables subject to the constraints Eq 9.9. These weights represent the combination of pixel values that are common within the positive class and significantly different in the negative class.

9.6 Support vector machines (SVM)

The first support vector machine algorithm was invented by Vapnik and Chervonenkis in the 1960s. Later, the ideas were developed further by various researchers in the 1990s. One of the earliest and simplest SVM models is the maximal margin classifier. To understand the main concepts, review the case of binary linear classification presented in Section 9.3.4. The optimization problem is defined by Eq 9.20. The norm of the weight vector is minimized; subject to the constraints, the product of the predicted and actual values of the output variable is greater than one for all the data points.

Optimization problems involving constraints are frequently solved by the Lagrangian method. This involves defining the Lagrangian, which is the sum of the objective function and a linear combination of all the constraints. The Lagrangian L for Eq 9.20 is written as follows:

$$\mathrm{L}(\mathbf{w},\mathrm{b}) = \frac{1}{2}\langle \boldsymbol{w}, \boldsymbol{w} \rangle - \sum_{i}^{l} \alpha_i \left(y_i \left(< \boldsymbol{w}, x_i > +b \right) - 1 \right) \qquad \textit{Eq 9.25}$$

This form is called the primal form of the optimization model. The first term on the right-hand side is half the norm of the weight vector, which is the objective function. (The factor half is used here simply for mathematical convenience; it does not affect the optimal values of variables.) The second part of the expression on the right-hand side is the linear combination of constraints. These represent the penalty for the violation of constraints. The terms α_i are called the Lagrange multipliers or Lagrange coefficients; these are non-negative constants that are used to combine all the constraints into a single function. If any point i is misclassified, the factor, $y_i(< \boldsymbol{w}, x_i > +b) - 1)$ is negative. Adding the negative of this quantity to the objective function causes L to become high. Thus, the linear combination of the constraint functions acts like a penalty function that penalizes points violating the constraints. Since we are minimizing the norm of the weight vector, the negative sign for the constraint terms ensures that the resulting function value is high in regions where constraints are violated. Because of this, the minimum of the Lagrangian is within the feasible region.

Exercise 9.6 Equipment breakdown classification – Lagrangian formulation

For the data in Exercise 9.3:

a) Write down the Lagrangian L in the primal form after substituting for the values of input and output variables.
b) How many Lagrange multipliers are present in the Lagrangian?

c) Verify that L does not depend on the coefficients of points that are on the constraint boundary.
d) Set the weight factors to zero. Then, find out the value of b that gives minimum misclassification.
e) Derive the conditions under which L is the minimum with respect to **w** and b.

Answer

a) Substituting the values of x_i and y_i in Eq 9.25,

$$L(\mathbf{w},b) = \frac{1}{2}(w_1^2 + w_2^2) +$$
$$\alpha_1(4w_1 + w_2 + b + 1) + \alpha_2(w_1 + b + 1)$$
$$-\alpha_3(6w_1 + b - 1) - \alpha_4(8w_1 + w_2 + b - 1)$$

b) Since there are four data points, there are four Lagrange multipliers.
c) The first input vector lies on the constraint boundary if $4w_1 + w2 + b = -1$. Since its output value y_i is negative and the resulting product $y_i(<\mathbf{w}, x_i> + b) = 1$, the term $(4w_1 + w_2 + b + 1)$ becomes 0. Therefore, the coefficient of α_1 in the earlier
d) Putting 0 for w_1, and w_2, the predicted output for all the points is b. Since two points are positive and two are negative, setting b to +1 or −1 will result in correct classification for one set of points, either positive or negative. The other two points will lie on the hyperplane and cause violation of the constraints.
e) The minimum of the Lagrangian with respect to **w** is obtained by setting the derivative to zero:

$$\frac{\partial L}{\partial w_1} = w_1 + 4\alpha_1 + \alpha_2 - 6\alpha_3 - 8\alpha_4 = 0$$

$$\frac{\partial L}{\partial w_2} = w_2 + \alpha_1 - \alpha_4 = 0$$

$$\frac{\partial L}{\partial b} = \alpha_1 + \alpha_2 - \alpha_3 - \alpha_4 = 0$$

These equations can be used to write the weight vector in terms of the Lagrange multipliers, and eliminate **w** from the primal form. Thus, L can be written in terms of α_i alone.

9.6.1 Dual form

The Lagrangian *L* contains additional variables α_i , apart from the original variables **w**. The optimal values of these variables must be found from the optimization theory. According to the theory, *L* should be minimized with respect to the original variables and maximized with respect to the Lagrange multipliers. This is called the Karush-Kuhn-Tucker (KKT) condition. According to this, the optimum is a saddle point obtained by maximizing with respect to α_i, the function obtained by minimizing *L* with respect to **w**. (See Figure 9.8 for an illustration of the saddle point). In other words, we look for the values of α_i for which L is the maximum at the optimal value of **w**. Intuitively, this can be understood as follows: If all the constraints are satisfied at any point, the term $y_i\left(\langle w, x_i\rangle + b\right) - 1$ is positive; therefore, the second term in the Lagrangian is negative if α_i is positive. The value of Lagrangian is reduced by using high positive values for α_i. This is avoided by maximizing L, which results in values for α_i shifting towards zero. On the other hand, at a point where constraints are violated, L is maximized by large values of α_i, which results in the weight factors moving towards regions having less penalty. Thus, the procedure of simultaneously maximizing with respect to α_i and minimizing with respect to **w** results in the selection of weight factors having the minimum norm and satisfying all the constraints.

By maximizing the Lagrangian, the penalty for the points that are misclassified is increased, and the coefficients of points that are correctly classified are set to zero.

The optimum satisfies the familiar condition that the partial derivatives of L with respect to **w** and **b** are zero. In addition, the following additional relationship known as the Kuhn-Tucker complementarity condition is obtained:

$$\alpha_i{}^*(y_i(< w^*, x_i > + b) - 1) = 0, \textit{ for all } i \qquad \textit{Eq 9.26}$$

Here, * in the superscript indicates the optimal values of the Lagrange multiplier and weight factors. This equation means that at the optimum, either the Lagrange multiplier is zero, or the constraint function is equal to zero. If $(y_i(< \mathbf{w}^*, x_i > +b) - 1)$ is greater than 0 for the optimal values of w, then the constraint is satisfied, and there is no need to add penalty for this point. Hence α_i is zero; zero is the value which maximizes the function $-\alpha_i^*(y_i(< \mathbf{w}^*, x_i > +b) - 1)$

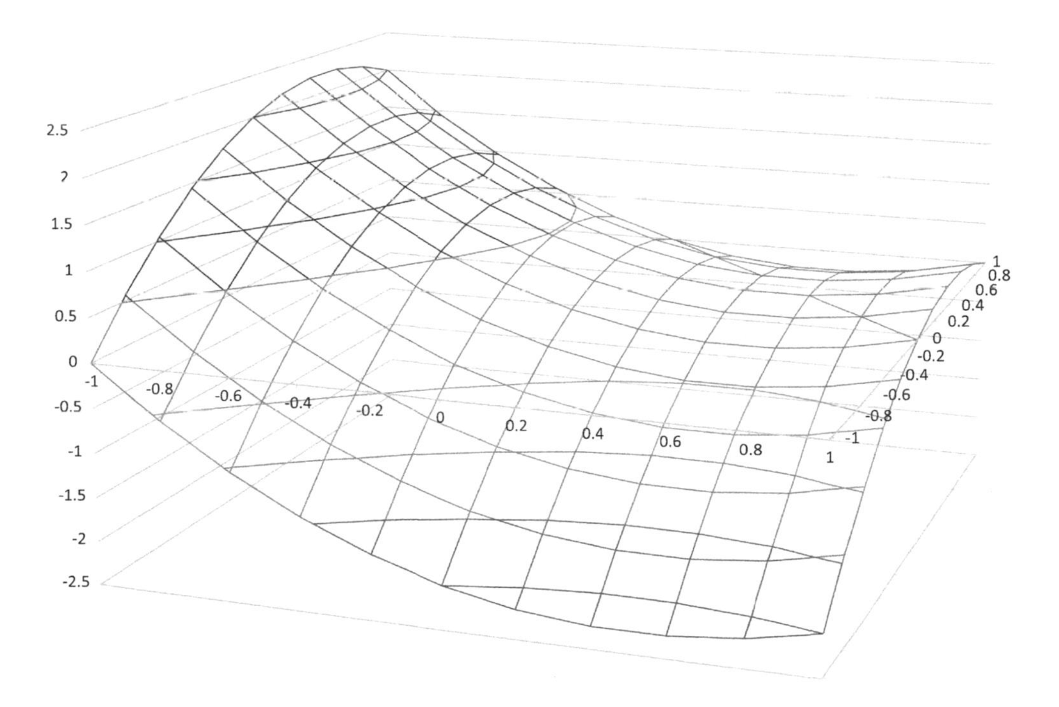

Figure 9.8 Saddle point

with respect to α_i because its coefficient in the Lagrangian is negative. If the Lagrange multiplier is zero, it means that the constraint is not active; the optimal point is within the interior of the feasible domain, not on the boundary or outside the boundary. The Lagrange multipliers should also satisfy

$$\alpha_i^* \geq 0 \qquad \textit{Eq 9.27}$$

Setting the derivatives of L with respect to w_j and b to zero, we get

$$\frac{\partial \mathrm{L}(\mathbf{w}, \mathrm{b})}{\partial w_j} = w_j - \sum_i^l \alpha_i y_i x_{i,j} = 0 \qquad \textit{Eq 9.28}$$

$$\frac{\partial \mathrm{L}(\mathbf{w}, \mathrm{b})}{\partial b} = \sum_i^l \alpha_i y_i = 0 \qquad \textit{Eq 9.29}$$

Eq 9.28 is used to eliminate the original optimization variable w_j from the optimization problem, using the substitution

$$w_j = \sum_i^l \alpha_i y_i x_{i,j} \qquad \textit{Eq 9.30}$$

This is written in vector notation as

$$\boldsymbol{w} = \sum_i^l \alpha_i y_i \boldsymbol{x_i} \qquad \textit{Eq 9.31}$$

The form of this equation is important. The weight vector is a linear combination of the input vectors! The Lagrange multipliers are the coefficients of the input vectors in this linear combination.

Substituting Eq 9.31 into Eq 9.25 and noting that the coefficient of b is zero because of Eq 9.29, the final expression is obtained as

$$\mathrm{L}(\boldsymbol{\alpha}) = \sum_i^l \alpha_i - \frac{1}{2} \sum_{i,j=1}^l y_i y_j \alpha_i \alpha_j < x_i, x_j > \qquad \textit{Eq 9.32}$$

This equation contains only the Lagrange multipliers. The original optimization variables are not present. The optimization problem is now written as

$$\mathrm{Maximize} \mathrm{L}(\boldsymbol{\alpha}) = \sum_i^l \alpha_i - \frac{1}{2} \sum_{i,j=1}^l y_i y_j \alpha_i \alpha_j < x_i, x_j >$$

Subject to

$$\sum_i^l \alpha_i y_i = 0, \text{ and, } \alpha_i \geq 0 \qquad \textit{Eq 9.33}$$

This model is called the *dual form* of the optimization problem. Note that the objective function is maximized, whereas in the primal form it is minimized. The original variables are not present in the dual form. The optimization problem has reduced to finding the coefficients α_i of the input points x_i, such that Eq 9.33 is maximized. These coefficients are either zero or positive, and accordingly they determine whether the point influences the decision boundary or not. This is because the weight factors that represent the decision boundary are a linear combination of the input points (Eq 9.31). The hyperplane is "supported" by a few input vectors. Hence the name support vector machine.

In the case of a linearly separable data set, it turns out that most of the coefficients are zero. Only a few coefficients are non-zero, corresponding to input vectors that lie closest to the hyperplane that discriminates the positive and negative classes. Due to Eq 9.33, the sum of coefficients of the positive points is equal to the sum of coefficients of the negative points. That is, support vectors from both classes have equal importance in determining the optimal hyperplane. The hyperplane is never determined by vectors belonging to one class alone.

The solution to the maximal margin classification problem is a hyperplane whose equation is a linear combination of the input vectors. Solving the dual form of the problem involves finding the coefficients of the input vectors in the equation to the hyperplane.

The dual form contains the term $< x_i, x_j >$, which is computed for each pair of input vectors. This term computes the dot product of a pair of vectors, a measure of how closely the two vectors are aligned, that is, the similarity between the two vectors. If two vectors belonging to the same class are very similar, the product $-y_i y_j < x_i, x_j >$ is a large negative number; therefore, the product of their coefficients $\alpha_i \alpha_j$ must be small to maximize L. Normally one of the coefficients become zero. Conversely, if two vectors from opposite classes are similar, their product of coefficients must be high. These vectors strongly determine the orientation of the maximal margin hyperplane. If each dot product $< x_i, x_j >$ is arranged in the row i and column j of a square matrix, the resulting matrix is called the Gram matrix, written as

$$G = \begin{bmatrix} < x_1, x_1 > & < x_1, x_2 > & \cdots & \cdots & < x_1, x_l > \\ \cdots & \cdots & < x_i, x_j > & \cdots & \cdots \\ < x_l, x_1 > & \cdots & \cdots & \cdots & < x_l, x_l > \end{bmatrix}$$

Exercise 9.7 Equipment breakdown classification – dual formulation

An example in one dimension is taken for the simplicity of illustration. For the data in Table 9.2:

a Compute the Gram matrix. Comment about the relative magnitudes of the elements in this matrix.

b Write down the Lagrangian in dual form. What is the nature of the objective function?

c Write down the constraints in the dual form. Interpret the meaning of the final expression.

d Calculate the solution to the dual problem by guessing the values, $\alpha_2 = 0; \alpha_4 = 0$. Check whether all the constraints in the primal problem are satisfied with this solution.

Answer

a The Gram matrix G is computed as follows:

16	4	24	32
4	1	6	8
24	6	36	48
32	8	48	64

The diagonal elements are nothing but the square of the norms of the vectors. The non-diagonal terms indicate the similarity between the corresponding vectors. Large magnitude means that the vectors are similar. For example, the third and fourth vectors show maximum similarity because G(3,4) = 48 is the highest value among non-diagonal elements. Both data points have high values for x1.

b The Lagrangian in dual form is a quadratic expression in α.

$$L(\alpha) = \alpha_1 + \alpha_2 + \alpha_3 + \alpha_4$$
$$-\frac{1}{2}(16\alpha_1^2 + \alpha_2^2 + 36\alpha_3^2 + 64\alpha_4^2)$$
$$-(4\alpha_1\alpha_2 - 24\alpha_1\alpha_3 - 32\alpha_1\alpha_4 - 6\alpha_2\alpha_3 - 8\alpha_2\alpha_4 + 48\alpha_3\alpha_4)$$

c The constraints are:

$$\alpha_1 \geq 0; \alpha_2 \geq 0; \alpha_3 \geq 0; \alpha_4 \geq 0$$

And

$$-\alpha_1 - \alpha_2 + \alpha_3 + \alpha_4 = 0$$

The sum of coefficients of the positive points is equal to the sum of coefficients of the negative points.

d Setting $\alpha_2 = 0$, $\alpha_4 = 0$, the constraints become:

$$\alpha_1 = \alpha_3$$

The Lagrangian reduces to:

$$L(\alpha) = 2\alpha_1 - 2\alpha_1^2$$

The maximum of L is found by setting the derivative to 0.

$$\frac{\partial L}{\partial \alpha_1} = 2 - 4\alpha_1 = 0$$

Therefore,

$$\alpha_1 = \alpha_3 = \frac{1}{2}$$

Points 1 and 3 have equal weightage in determining the optimal hyperplane.

$$w = \sum_{i}^{l} \alpha_i y_i \boldsymbol{x}_i = \frac{(\boldsymbol{x}_3 - \boldsymbol{x}_1)}{2} = [1]$$

b is calculated by noting that the points 1 and 3 have a functional margin = 1, that is,

$$< \boldsymbol{w}, \boldsymbol{x}_1 > +b = -1; < \boldsymbol{w}, \boldsymbol{x}_3 > +b = 1; \rightarrow 10w + 2b = 0 \rightarrow b = -5$$

Therefore, the discriminant has the equation

$$x1 - 5 = 0$$

With this hyperplane, the actual and predicted values for all the points are tabulated in Table 9.8:

The predicted value has the same sign as the actual class label. In addition, the functional margin is greater than or equal to 1. Therefore, the trial solution $\boldsymbol{\alpha}$ = [0, 0.5, 0.5, 0] is the actual solution to the dual problem. The points 1 and 3 are the support vectors having non-zero Lagrange multipliers.

Exercise 9.8 Equipment breakdown classification 2 – dual formulation

This example involves two input variables. Using the data in Table 9.4:

a Compute the Gram matrix.
b Write down the Lagrangian in dual form. What is the nature of the objective function?
c Write down the constraints in the dual form.
d Write down the expression for the weight vector using the actual values of input variables.
e Plot the points and visually locate the maximal margin hyperplane

Answer

a The Gram matrix G is computed as follows:

$$\begin{matrix} 17 & 4 & 24 & 33 \\ 4 & 1 & 6 & 8 \\ 24 & 6 & 36 & 48 \\ 33 & 8 & 48 & 65 \end{matrix}$$

The Gram matrix is very similar to the previous exercise, because the second variable x2 has low values.

Table 9.8 Equipment Breakdown Classification – Predicted Class Label

i	*x1*	*y (actual)*	*y (predicted)*
1	4	−1	−1
2	1	−1	−2
3	6	1	1
4	8	1	3

b The Lagrangian in dual form is a quadratic expression in α.

$$L(\alpha) = \alpha_1 + \alpha_2 + \alpha_3 + \alpha_4$$

$$-\frac{1}{2}(17\alpha_1^2 + \alpha_2^2 + 36\alpha_3^2 + 65\alpha_4^2)$$

$$-(4\alpha_1\alpha_2 - 24\alpha_1\alpha_3 - 33\alpha_1\alpha_4 - 6\alpha_2\alpha_3 - 8\alpha_2\alpha_4 + 48\alpha_3\alpha_4)$$

c The constraints are:

$$\alpha_1 \geq 0;\ \alpha_2 \geq 0;\ \alpha_3 \geq 0;\ \alpha_4 \geq 0$$

And

$$-\alpha_1 - \alpha_2 + \alpha_3 + \alpha_4 = 0;$$

The constraints are the same as that of the previous exercise.

d The weight vector is calculated as

$$\begin{bmatrix} w_1 \\ w_2 \end{bmatrix} = -\alpha_1 \begin{bmatrix} 4 \\ 1 \end{bmatrix} - \alpha_2 \begin{bmatrix} 1 \\ 0 \end{bmatrix} + \alpha_3 \begin{bmatrix} 6 \\ 0 \end{bmatrix} + \alpha_4 \begin{bmatrix} 8 \\ 1 \end{bmatrix}$$

e The points are plotted in Figure 9.9. Since the maximal margin hyperplane is equidistant from the nearest positive and negative points, the line labeled H in the figure is its likely position. Points P1, P3, and P4 are the support vectors and have a functional margin = 1. These points have non-zero Lagrange multipliers.

9.6.2 Learning with a few data points in a high dimensional space

In many problems, there are thousands of input variables, but very few data points. In image recognition, if the images are of size 100 × 100 pixels, there are 10,000 variables that represent all the pixel values. If there are only two images in the training set, one for the positive class and one fore negative class, is it possible to learn the discriminant that separates the two classes? Let us see what happens when we maximize the Lagrangian when there are only two data points **P1** and **P2**, where **P1** is positive and **P2** is negative. The Lagrangian is written as:

$$L(\alpha) = \alpha_1 + \alpha_2 - \frac{1}{2}(\alpha_1^2 P_1^2 + \alpha_2^2 P_2^2 - 2\alpha_1\alpha_2 \vec{P_1}.\vec{P_2}$$

From Eq 9.33, the Lagrange coefficients for the positive and negative points are equal ($\alpha_1 = \alpha_2$). Therefore,

$$L(\alpha) = 2\alpha_1 - \frac{1}{2}\left(\alpha_1^2 p_1^2 + \alpha_1^2 p_2^2 - 2\alpha_1^2 \vec{P_1}.\vec{P_2}\right) = 2\alpha_1 - \frac{\alpha_1^2}{2}\left(P_1^2 + P_2^2 - 2\vec{P_1}.\vec{P_2}\right)$$

The coefficient of the quadratic term in the brackets is nothing but the square of the length of the vector $(\vec{P_1} - \vec{P_2})$. Therefore,

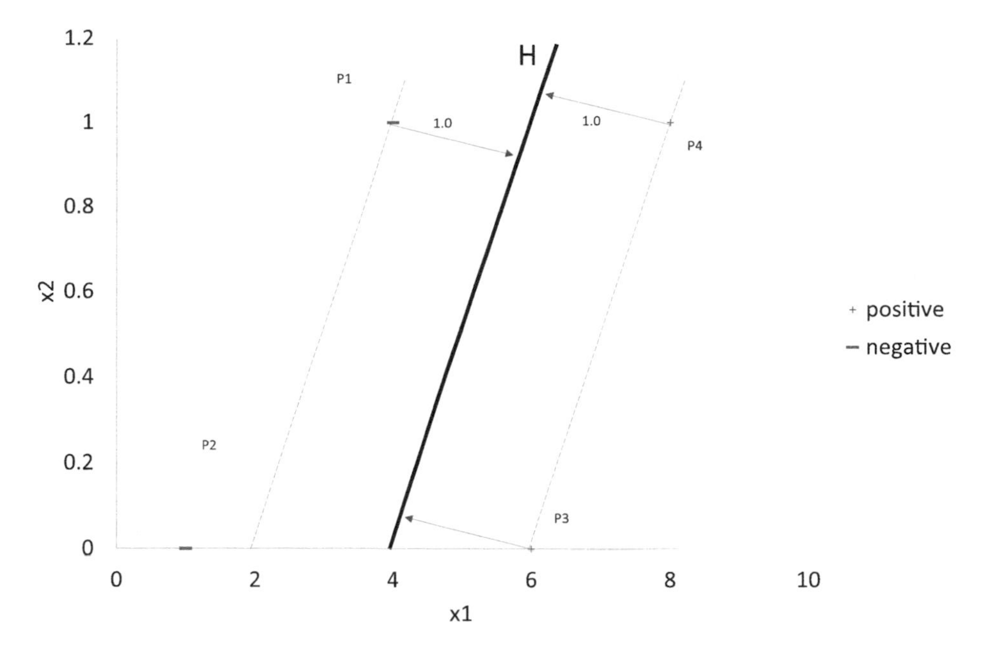

Figure 9.9 The decision boundary is equidistant from the nearest positive and negative points

$$L(\alpha)=2\alpha_1-\frac{\alpha_1^{2}}{2}\left\|\vec{P_1}-\vec{P_2}\right\|^2$$

Setting the derivative with respect to the Lagrange coefficient to zero,

$$2-\alpha_1\left\|\vec{P_1}-\vec{P_2}\right\|^2=0$$

Therefore,

$$\alpha_1=\frac{2}{\left\|\vec{P_1}-\vec{P_2}\right\|^2}$$

The weight vector is a linear combination of the support vectors as follows:

$$w=\alpha_1\left(\vec{P_1}-\vec{P_2}\right)=\frac{2}{\left\|\vec{P_1}-\vec{P_2}\right\|^2}\left(\vec{P_1}-\vec{P_2}\right)$$

The weight vector is in the direction of the unit vector from P1 to P2. The bias term is calculated using Eq 9.21.

$$b=-\left(\frac{1}{2}\right)\left[\left\langle w,\vec{P_1}\right\rangle+\left\langle w,\vec{P_2}\right\rangle\right]=-\left(\frac{1}{2}\right)\frac{2}{\left\|\vec{P_1}-\vec{P_2}\right\|^2}\left(\vec{P_1}-\vec{P_2}\right).\left[\vec{P_1}+\vec{P_2}\right]=\frac{p_1^{\,2}-p_2^{\,2}}{\left\|\vec{P_1}-\vec{P_2}\right\|^2}$$

The predicted value of the output is obtained as

$$yp = \langle w, x \rangle + b = \frac{2}{(\overrightarrow{P_1} - \overrightarrow{P_2})^2} (\overrightarrow{P_1} - \overrightarrow{P_2}).\vec{x} + b$$

The first term in the previous expression computes the projection of x in the direction of $(\overrightarrow{P_1} - \overrightarrow{P_2})$. To understand what is predicted by the expression, assume that the origin is at the point P2, that is, P2 = 0 and b = −1. Then after substituting for the value of b, yp is written as

$$yp = \frac{2}{\overrightarrow{P_1}^2} (\overrightarrow{P_1}).\vec{x} - 1$$

If x is perfectly aligned with $\overrightarrow{P_1}$, the predicted value is 1; if x is zero, the predicted value is −1. If x is exactly midway between 0 and $\overrightarrow{P_1}$, the predicted value is 0. That is, the prediction depends on how far away the point is from $\overrightarrow{P_1}$ and $\overrightarrow{P_2}$. The prediction depends on the output value of the nearest support vector. Therefore, it is similar to the prediction from the *nearest neighbor method*, in which the output values of the nearest data points are used for the prediction. Even if there are only two data points, the maximal margin classification is able to predict the class label using a method similar to nearest neighbor approach! The entire space is divided into two halves, midway between the positive and negative data points. The half space on the side of the positive points has a positive prediction, and the other half has a negative prediction.

If another data point is added to the training data, there will be three support vectors in general. Assume the third data point P3 is positive. Then, the hyperplane should satisfy the following equations:

$$\langle w, \overrightarrow{P_1} \rangle + b = 1$$

$$\langle w, \overrightarrow{P_2} \rangle + b = -1$$

$$\langle w, \overrightarrow{P_3} \rangle + b = 1$$

Here, we have three equations in (n + 1) variables, where n is the number of input variables. Even though in general there are many possible solutions for w and b, the solutions should satisfy certain conditions. Subtracting the equations for the positive points,

$$\langle w, \overrightarrow{P_1} - \overrightarrow{P_3} \rangle = 0$$

That is, the weight vector should be normal to the vectors connecting every pair of positive points and every pair of negative points that are support vectors. That is, all the positive support vectors lie on a plane that is perpendicular to the weight vector. Similarly, all the negative support vectors lie on another plane lying perpendicular to the weight vector. (These two planes are separated by a distance equal to two divided by the length of the weight vector.) Since the weight vector contains *n* unknown coefficients, at most (n + 1) distinct data points can satisfy the conditions $\langle w,x \rangle + b = \pm 1$. That is, there can be a maximum of (n + 1) support vectors. However, even with just two data points of the opposite classes, the classifier can learn a hyperplane that separates the two classes. The addition of each new data point would cause the hyperplane to rotate such that all the nearest data points have a functional margin of 1.

9.6.3 Important characteristics of the problem formulation

The equation for the maximal margin hyperplane is written as a linear combination of support vectors. Usually, only a few data points are support vectors, which are the points that lie closest to the hyperplane. All the remaining points have Lagrange multipliers equal to zero. This property is called the sparseness of the solution. Due to this property, all the data points that are not support vectors can be removed from the gram matrix, and the solution is not changed. The support vectors contain all the information needed to construct the hyperplane. It has also been shown that the generalization capability is higher as the number of support vectors becomes small. This again supports the Occam's principle of selecting the simplest model. Another implication of the sparseness of the solution is that we do not need too many data points to learn the maximal margin hyperplane. Even if there are more than one thousand variables, you may need only two or three data points to learn the separation between positive and negative classes. The weight factors for all one thousand variables are determined by the two or three support vectors. The data points that are far from the decision surface are discarded, and even if they are missing in the data set, the discriminant is not affected. The sparseness of the solution is important in selecting good techniques for solving the problem. This will be discussed in the next section.

The maximal margin formulation works only in the case of linearly separable data sets. In the case of non-separable data sets, there are no feasible solutions in the primal formulation. For the dual formulation, the objective function is unbounded and will not converge. In such cases, learning must be performed in non-linear feature spaces, or models such as soft margin classifiers (Section 9.7) should be used.

9.6.4 Solving the dual form

The dual form is a quadratic optimization problem subject to linear constraints. For such problems, it can be shown that the local optimum is the same as the global optimum. That is, multiple local optima are not present. This permits the use of local search techniques which are very efficient, and you are guaranteed to obtain the globally optimal solution. The simplest method is gradient descent since the dual form is easily differentiable. However, the equality constraint resulting from the KKT condition (Eq 9.33) should be satisfied. That is, each step in the downhill direction should be taken such that the sum of Lagrange multipliers for the positive points should be equal to the sum for negative points. Furthermore, since there are as many optimization variables as the number of data points, the sparseness of the solution should be effectively exploited to solve large problems involving thousands of data points. An efficient algorithm that makes use of these properties is Sequential Minimal Optimization (SMO). See Cristianini and Taylor (2000) for details.

9.6.5 Prediction

After solving the dual problem given by Eq 9.33, the Lagrange multipliers are obtained. Using these coefficients, the weight vector is computed using Eq 9.31. This defines the hyperplane that separates the positive and negative points. A new data point $\mathbf{x}_t$ (test point) is classified by evaluating the distance of the point from the hyperplane y_p using the equation,

$$y_p = \langle \mathbf{w}, \mathbf{x}_t \rangle + b \qquad \text{Eq 9.34}$$

This equation gives the prediction of the classifier for the test point. The sign of y_p determines the class of the point, whether it is positive or negative. If y_p is between −1 and +1, the functional margin of the test point is less than 1; it is closer to the hyperplane than the support vectors. The probability of misclassification is higher when the margin is close to zero.

Substituting the value of **w** from Eq 9.31, Eq 9.36 is written as

$$y_p = \langle \mathbf{w}, \mathbf{x}_t \rangle + b = \sum_i^l \alpha_i y_i \langle \mathbf{x}_i, \mathbf{x}_t \rangle + b = \sum_{i \in sv} \alpha_i y_i \langle \mathbf{x}_i, \mathbf{x}_t \rangle + b \qquad \text{Eq 9.35}$$

Where $i \in sv$ indicates that the summation needs to be carried out only for the support vectors. (The other coefficients are zero.) The prediction for a data point $\mathbf{x}_t$ depends only on its dot product with the support vectors, that is, its similarity with the support vectors. This is important when extending the algorithm to non-linear classification as will be discussed in Section 9.6.7. Note that the dual objective function as well as the expression for b (Eq 9.21) involves dot products of vectors and not the vectors themselves.

9.6.6 Kernels

The concept of learning in the feature space was introduced in 9.5.1. Kernel is a mathematical tool for learning in the feature space without explicitly computing the features. Kernel is a function that computes the dot product of two feature vectors as follows:

$$K(x_i, x_j) = \langle \varphi(x_i), \varphi(x_j) \rangle \qquad \text{Eq 9.36}$$

The simplest kernel is the dot product of two vectors.

$$K(x_i, x_j) = \langle x_i, x_j \rangle \qquad \text{Eq 9.37}$$

Here, the features are the original vectors themselves and represents a linear kernel. A more generalized form of the linear kernel involves a linear transformation of the vector using a matrix **A** as follows:

$$\varphi(x_i) = [A]\, x_i$$

This feature vector represents a rotation or scaling of the vector x_i. The kernel function now becomes

$$K(x_i, x_j) = \langle [A]x_i, [A]x_j \rangle = x_i^T [A]^T [A] x_j = x_i^T [B] x_j \qquad \text{Eq 9.38}$$

where the superscript T denotes the transpose of the matrix, and B is a symmetric square matrix obtained as the product of $[A]^T [A]$. From linear algebra, this matrix B is positive semi-definite. Note that the last part of Eq 9.38 permits us to compute the kernel without evaluating the feature vectors explicitly. Any positive semi-definite symmetric square matrix B could be used to compute the linear kernel.

A simple example of a non-linear kernel is a dot product of feature vectors involving monomials of degree two. Consider a two-dimensional input vector as follows:

$$\mathbf{x}_i = \begin{bmatrix} x_{i,1} \\ x_{i,2} \end{bmatrix}$$

If the feature vector is written as follows,

$$\varphi(\boldsymbol{x}_i) = \begin{bmatrix} x_{i,1}^{\ 2} \\ 2x_{i,1} \times x_{i,2} \\ x_{i,2}^{\ 2} \end{bmatrix}$$

The dot product of two feature vectors $\varphi(x_i), \varphi(x_j)$ is

$$\langle \varphi(x_i), \varphi(x_j) \rangle = x_{i,1}^{\ 2} x_{j,1}^{\ 2} + 2x_{i,1}x_{i,2}x_{j,1}x_{j,2} + x_{i,2}^{\ 2} x_{j,2}^{\ 2} = \left(x_{i,1}x_{j,1} + x_{i,2}x_{j,2}\right)^2$$
$$= \left(\langle x_i, x_j \rangle\right)^2$$

Therefore, the kernel consisting of these features is simply the square of the simple linear kernel. A polynomial kernel of any order m can be constructed by taking the power of the linear kernel as follows:

$$K(x_i, x_j) = \left(\langle x_i, x_j \rangle\right)^m \qquad \textit{Eq 9.39}$$

This is interesting because these kernels are simply more complex functions of simple kernels. Many kernels are created this way. However, all kernels are not like this. A popular kernel used for non-linear learning is a Gaussian kernel, also known as the radial basis function. It is defined as follows:

$$K(x_i, x_j) = e^{-\frac{x_i - x_j^{\ 2}}{2\sigma^2}} \qquad \textit{Eq 9.40}$$

The exponent in the expression contains the normalized distance between the two vectors $\mathbf{x}_i$ and $\mathbf{x}_j$. Two vectors that are identical have the maximum value of 1 for the kernel function. As the distance between the vectors increases, the kernel function drops to zero (See Figure 9.10). Points that are at the same distance have the same value of the function. Therefore, this function represents our intuitive understanding of similarity between two vectors.

9.6.7 Non-linear classification using kernels

If the feature vector is used for classification instead of the original input vector, wherever the dot product of input vectors appears in the learning algorithm we can replace it with the dot product of the feature vector. Since the dot product of the feature vector is defined as the kernel, extension of the support vector algorithms for the non-linear case is straightforward – simply replace the dot products of input vectors with the kernel function. Thus, equations Eq 9.32, Eq 9.35, and Eq 9.21 become

$$\mathrm{L}(\alpha) = \sum_{i}^{l} \alpha_i - \frac{1}{2} \sum_{i,j=1}^{l} y_i y_j \alpha_i \alpha_j K(x_i, x_j) \qquad \textit{Eq 9.41}$$

$$y_p = \sum_{i \in \mathrm{sv}} \alpha_i y_i K(\boldsymbol{x}_i, \boldsymbol{x}_t) + \mathrm{b} \qquad \textit{Eq 9.42}$$

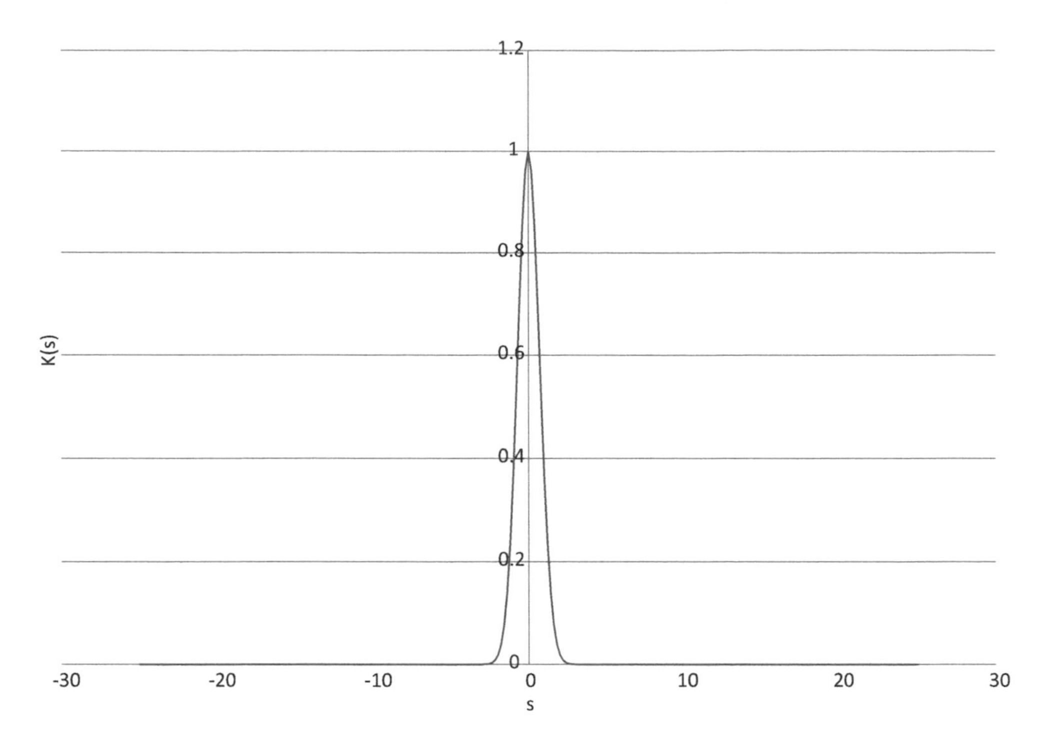

Figure 9.10 Radial basis function in terms of the distance between points s

$$b = -\left(\frac{1}{2}\right)\left[\min\left(\sum_{i \in sv} \alpha_i y_i \mathrm{K}\left(\boldsymbol{x}_i, \mathbf{x}^{+}\right)\right) + \max\left(\sum_{i \in sv} \alpha_i y_i \mathrm{K}\left(\boldsymbol{x}_i, \mathbf{x}^{-}\right)\right)\right] \qquad \textit{Eq 9.43}$$

As in the linear case, learning involves finding the coefficients such that the Lagrangian in the feature space (Eq 9.42) is the maximum. Using these coefficients, the classes are predicted using Eq 9.42 which also involves the Kernel function.

If non-linear kernels are used for learning, the resulting decision boundary will be non-linear in the original input space. For example, the decision boundary obtained using a Gaussian kernel is

$$\sum_{i \in \mathrm{sv}} \alpha_i y_i e^{-\frac{x - x_i^2}{2\sigma^2}} + \mathrm{b} = 0 \qquad \textit{Eq 9.44}$$

Exercise 9.9 Gaussian kernel with two support vectors

A Gaussian kernel with $\sigma^2 = 0.1$ was used in support vector classification. There were two support vectors; $x_1 = [1,0]$ belonging to the positive class, and $x_2 = [-1,0]$ belonging to the negative class. What is the shape of the decision boundary? Derive the expression that predicts the class for a test point. Plot the regions where reliable prediction is possible.

Answer

From KKT complementarity conditions, the Lagrange multipliers are equal, that is, $\alpha_1 = \alpha_2$. The decision boundary is given by:

$$\alpha_1 e^{-\frac{\|x-x_1\|^2}{2\sigma^2}} - \alpha_1 e^{-\frac{\|x-x_2\|^2}{2\sigma^2}} + b = 0$$

The support vectors have a margin of 1. Therefore

$$\alpha_1 e^{-\frac{\|x_1-x_1\|^2}{2\sigma^2}} - \alpha_1 e^{-\frac{\|x_1-x_2\|^2}{2\sigma^2}} + b = 1$$

That is,

$$\alpha_1 \left(1 - e^{-10}\right) + b = 1$$

Similarly, for the second point,

$$\alpha_1 \left(e^{-10} - 1\right) + b = -1$$

From the two previous equations, we get b = 0, and α_1 is approximately equal to 1. Therefore, the equation for the decision boundary is:

$$e^{-\frac{\|x-x_1\|^2}{2\sigma^2}} = e^{-\frac{\|x-x_2\|^2}{2\sigma^2}}$$

The equation is satisfied by the line $\mathbf{x} = [0,0]$. Therefore, the decision boundary is a straight line passing through the origin.

The prediction of the model is given by the Eq 9.45:

$$y_p - \sum_{i \in sv} \alpha_i y_i K(\mathbf{x}_i, \mathbf{x}_t) + b = \alpha_1 \left(e^{-\frac{\|x_t-x_1\|^2}{2\sigma^2}} - e^{-\frac{\|x_t-x_2\|^2}{2\sigma^2}} \right) \qquad \text{Eq 9.45}$$

Points close to the decision boundary have a low margin, and the predictions are not reliable. The probability of misclassification is high. For reliable prediction, the margin should be greater than or equal to one. Points having a margin close to one are near the support vectors, and those having a margin higher than 1 are likely to be within the region covered by other points in the training data set. These are regions where reasonably accurate predictions are possible.

From Eq 9.45, when $\boldsymbol{x}_t$ is close to $\boldsymbol{x}_1$, the first exponent becomes close to zero, and the first term becomes close to one; the resultant is a positive value. On the contrary, if $\boldsymbol{x}_t$ is close to $\boldsymbol{x}_2$, the second term becomes one, and the resultant prediction is negative. Thus, the points close to the first support vector are predicted to be positive, and the points close to the second support vector are predicted to be negative. Thus, the radial basis function kernel resembles the nearest neighbor method in the prediction. Curves that predict the same values are ellipses centered around the support vectors. Since σ^2 is small, the region having a margin of 1 is a tiny ellipse, too small to be visible. Hence a higher value of $\sigma^2 = 1$ is selected for illustration. This value results in the value of the Lagrange multiplier $\alpha_1 = 1.582$. In Figure 9.11, randomly generated points having margin greater than 0.9 are plotted. There are two clusters of points, one around the positive support vector and the other around the negative support vector. Note that the support vectors are not exactly at the center of the clusters because of the contribution from the opposite point in Eq 9.45. The regions where there are no points plotted have ambiguity in predictions because no training data lies in this region. However, if test data points lie in this region, we classify them as positive or negative depending on the sign of the predicted value. In this example, the decision boundary is the vertical axis at the origin. All the points lying on the left side are negative and vice versa.

Exercise 9.10 Gaussian kernel with three support vectors

A Gaussian kernel with $\sigma^2 = 1$ was used in support vector classification. There were three support vectors: $x_1 = [0,0]$ belonging to the positive class; and $x_2 = [-1,0]$, $x_3 = [0, -1]$ belonging to the negative class. What is the shape of the decision boundary? Derive the expression that predicts the class for a test point. Plot the regions where reliable prediction is possible.

Answer

Since the negative support vectors have the same distance from the positive support vector, both the negative points are equivalent. Therefore, $\alpha_2 = \alpha_3$. From KKT complementarity conditions, the Lagrange multipliers satisfy the relationship, $\alpha_1 = 2\alpha_2$. The class prediction is given by:

$$y_p = 2\alpha_2 e^{-\frac{x_t - x_1^2}{2\sigma^2}} - \alpha_2 e^{-\frac{x_t - x_2^2}{2\sigma^2}} - \alpha_2 e^{-\frac{x_t - x_3^2}{2\sigma^2}} + b \qquad \text{Eq 9.46}$$

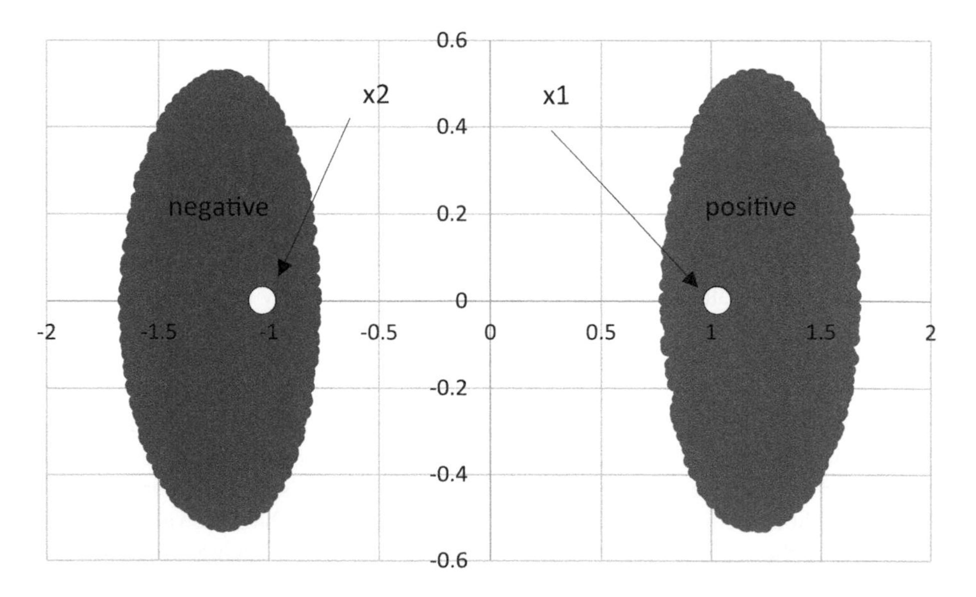

Figure 9.11 The decision boundary defined by a Gaussian kernel with two support vectors

The support vectors have a margin of 1. Substituting +1 and −1 for y_p for the first two support vectors, we get two equations in two unknown variables α_2 and b. Solving,

$$\alpha_2 = 2.124 \text{ and } b = -0.671$$

Like the previous example, test points close to the positive support vector x1 will be predicted as positive since the first term in Eq 9.46 will be large; the negative terms will become small as the distance between the test point and the negative support vectors increase. In order to visualize the decision boundary, points were randomly generated and predicted using Eq 9.46. Points having a margin between 0.9 and 1.1 are plotted in Figure 9.12. The decision boundary with zero margin is between the edges of the positive and negative clusters shown.

In this example, the positive support vector has a small region of influence, indicated by the points having positive predictions with high margin. This region is an ellipse containing the positive support vector. The region of influence of the negative support vectors overlap to form a contiguous area having a complex non-linear boundary. As σ is decreased, the region of influence decreases, and the space is partitioned into multiple disconnected regions around each support vector. In Figure 9.13, randomly generated points having a margin between 0.9 and 1.1 are plotted for the Gaussian kernel with $\sigma^2 = 0.1$.

This example brings out important characteristics of the Gaussian kernel. The nature of the decision surface changes dramatically as the model parameter σ^2 is changed. As this parameter is reduced, the space gets fragmented into more and more chunks, and the number of support vectors increase. While this might increase the training accuracy, generalization errors are likely to increase because of overfitting.

Non-linear kernels with tunable parameters might show good training accuracy but have poor generalizability and result in high prediction errors on unseen data. They tend to suffer from overfitting.

9.6.8 Selecting kernels for an application

Kernels have to be carefully chosen such that the features implicitly represented by them are able to separate the classes. Thus, kernels represent key domain knowledge. Even though, there is considerable flexibility in defining the kernel functions, the kernel matrices should satisfy these properties:

- Symmetric
- Positive semi-definite

The requirement that the kernel matrix is symmetric means that the order of input vectors should not affect the value of the kernel function. That is, the similarity between vectors x_i and x_j should be the same as that between x_j and x_i.

A positive semi-definite matrix is one whose eigenvalues are non-negative. The requirement that the kernel matrix should be positive semi-definite comes from a mathematical theorem called Mercer's theorem. This condition is essential for the convergence of the solutions.

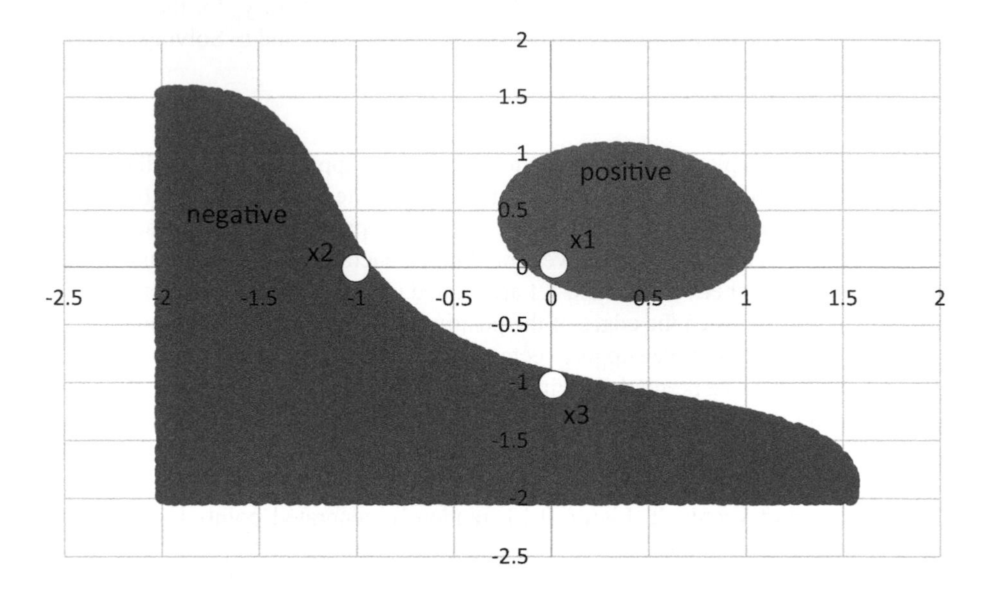

Figure 9.12 The decision boundary defined by a Gaussian kernel with three support vectors

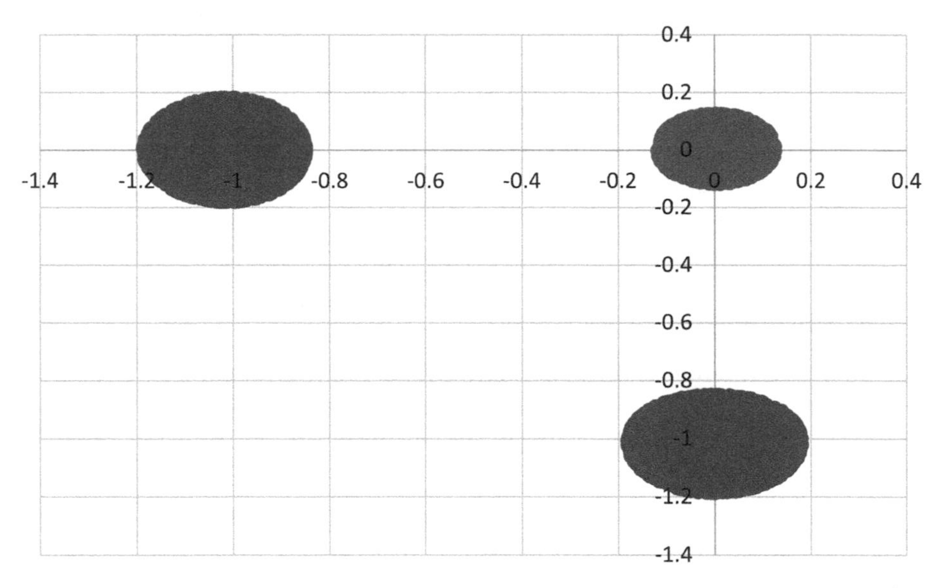

Figure 9.13 The decision boundary defined by a Gaussian kernel with three support vectors, $\sigma^2 = 0.1$.

9.7 Soft margin classification

When data are not linearly separable, it might be tempting to use complex non-linear kernels. However, as shown in the previous section, this might result in overfitting. Another option is to use a soft margin classifier. This approach eliminates problems due to noise and outliers in data that make the problem not linearly separable; whatever hyperplane is chosen, some points will always be misclassified because of noise. In such situations, the maximal margin learning algorithm discussed in Section 9.3.4 (also known as hard margin classification) does not converge.

The main concept behind soft margin classification is that a certain amount of misclassification is permitted, that is, margin constraints are allowed to be violated to a certain degree. This is done by introducing slack variables ξ_i in the optimization model as follows:

$$\left(y_i\left(\langle w, x_i\rangle + b\right)\right) \geq 1 - \xi_i \qquad \textit{Eq 9.47}$$

Instead of strictly enforcing the margin to be greater than or equal to one, a small reduction up to ξ_i is permitted, for each point i. The slack variable should be greater than or equal to 0. When it is zero, the margin constraint is satisfied perfectly. The higher the value of ξ_i, the more relaxation in the constraint is allowed. Instead of arbitrarily selecting the value of ξ_i, the slack variables are optimized such that there is a trade-off between the maximum margin and the degree of violation. In general, we want to minimize the slack along with the norm of the weight vector. (Remember, the norm of the weight vector determines the margin.) There are different ways of formulating the optimization problem to achieve this. One method is called the *2-norm soft margin classification* in which the sum of squares of the slack variables are minimized as follows:

$$\mathrm{O}(\mathrm{w}, \mathrm{b}, \xi_i) = \frac{1}{2}\langle w, w\rangle + \frac{C}{2}\sum_{i}^{l}\xi_i^{\,2} \qquad \textit{Eq 9.48}$$

C is a hyperparameter of the model, which controls the balance between the margin and slack. The higher the value of C, the larger is the penalty for margin violation. The parameter C should be chosen a-priori. In practice, different values of C are evaluated using validation data, and an appropriate number is chosen for the given data set.

Another method is called *1-norm soft margin classification* in which the objective function is slightly different:

$$\mathrm{O}(\mathrm{w}, \mathrm{b}, \xi_i) = \frac{1}{2}\langle w, w\rangle + C\sum_{i}^{l}\xi_i \qquad \textit{Eq 9.49}$$

In either case, the approach is to formulate the Lagrangian and derive the dual problem by setting the derivative to zero and imposing thc stationarity conditions. The derivation is given in Cristianini and Tayler (2000). The final expressions for the 1-norm soft margin case are given here.

The primal Lagrangian:

$$\mathrm{L}(\mathrm{w}, \mathrm{b}, \xi_i) = \frac{1}{2}\langle w, w\rangle + C\sum_{i}^{l}\xi_i - \sum_{i}^{l}\alpha_i\left(y_i\left(\langle w, x_i\rangle + b\right) - 1 + \xi_i\right) - \sum_{i}^{l} r_i\xi_i \qquad \textit{Eq 9.50}$$

Where r_i are new Lagrange multipliers to ensure that $\xi_i \geq 0$

The dual form is obtained as follows:

$$\text{Maximize}\,\mathrm{L}(\alpha) = \sum_{i}^{l}\alpha_i - \frac{1}{2}\sum_{i,j=1}^{l} y_i y_j \alpha_i \alpha_j \langle x_i, x_j\rangle$$

Subject to

$$\sum_{i}^{l} \alpha_i y_i = 0 \,,\, \alpha_i \geq 0 \text{ and, } \alpha_i \leq C \qquad Eq\ 9.51$$

Comparing with Eq 9.33, the only difference with the hard margin case is that there is an additional constraint on α_i, that is, $\alpha_i \leq C$. This is called a box constraint, since the value of α should lie within the box defined by [0,C]. The Lagrange multipliers are bounded by the value of C chosen for the problem. Without this condition, the value of α_i becomes arbitrarily large when the data set is not linearly separable.

Since the 1-norm soft margin case is very similar to the hard margin case, the solution techniques that were discussed earlier are applicable. During the updating of the Lagrange multipliers, it should be ensured that the values do not exceed C.

9.8 Deep learning

It was emphasized in the previous sections that domain specific features are essential for learning the correct discriminant. For example, in the image recognition task, the Prewitt mask was used to extract horizontal and vertical edges in images; these were used to separate images into classes. Other masks (filters) such as the Robinson mask have been developed to detect edges that are inclined. If inclined edges are important for separating images into classes, then these masks should be used. These masks differ from the Prewitt mask in the values of coefficients in their matrices. This raises the question: can the mask itself be learnt, instead of hardcoding the matrices? In other words, can features be learnt from data? This is the concept behind deep learning.

Deep learning attempts to discover features by treating feature vectors as unknown variables. Deep learning networks are typically artificial neural networks containing many layers. The initial layers are meant to learn the features that discriminate the classes; the final layers represent the relationship between features that separates the classes.

A convolution neural network (CNN) is a type of deep learning network. It consists of three types of layers:

a Convolution layer
b Pooling layer
c Fully connected layer

Each layer takes an input volume and transforms it to create an output volume. The output volume of one layer becomes the input volume for the next layer. In general, the input and output volumes are 3-dimensional (width, height, and depth). In the case of image data, the input volume contains the pixel values, which are arranged in the width and height dimensions. The color values or channels form the depth layer.

The convolution layer is schematically shown in Figure 9.14. The transformation is performed by multiplying by a matrix of weights. The weights are the equivalent of the Prewitt mask; however, the weights in a CNN are unknown and determined through optimization. The weight matrix is of size much smaller than the input volume. For example, the weight matrix might be of dimensions 3 × 3 while the image might be of size 256 × 256. The same set of weights is used to extract features from the entire image by selecting windows within the images and moving the windows horizontally and vertically. The values within

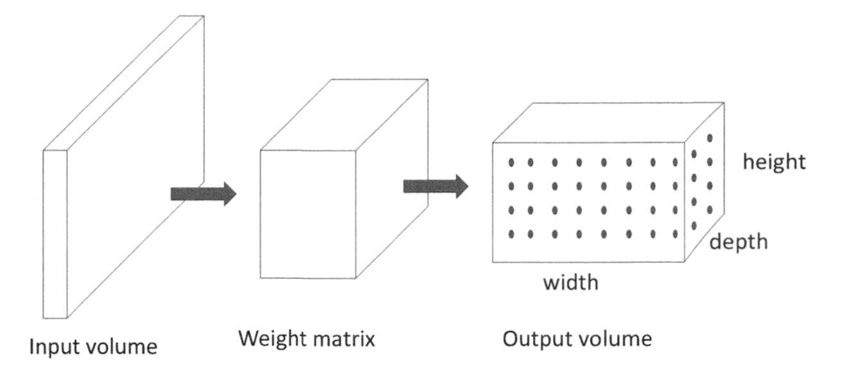

Figure 9.14 Convolution layer

the selected window are multiplied with the weight matrix and summed up; this is similar to taking the dot product of two matrices. This operation is repeated for all the windows by moving from left to right and from top to bottom. Convolution is the process of moving the windows in this manner and performing the dot product. The weight matrix is said to convolve in the horizontal and vertical directions. At each position of the window, a value is obtained, resulting in a feature map indicating the presence of the feature at each position.

It is not necessary to use the entire feature map for learning. It is usually enough to determine whether there is a feature within a certain part of the image. Therefore, typical values of features are extracted using *down sampling*, selecting a few values from the feature map. This is done by the pooling layer. Commonly used functions for pooling are MAX and MIN. MAX pooling involves selecting the maximum value from a region of the feature map. MIN pooling selects the minimum value. The size of the output volume is reduced by using these operators. By taking maximum or minimum values, the output does not change if the location of the features is changed by a few pixels. For example, if a horizontal edge is shifted by a single pixel, the output from the pooling layer will be the same.

The fully connected layer resembles a standard neural network in which each node in the input layer is connected to each node in the output layer. The output functions are standard ones used in ANN. Support vector machines have also been used for the final layer that computes the class labels in classification problems.

The convolution network ensures that the number of weight factors used for learning is kept to the minimum. Since the same set of weights in the convolution layer is repeated for the entire image, the number of weights used to extract features is significantly lower than a conventional neural network that requires one weight factor for each connecting edge.

CNN has been successful in image recognition as well as learning from time series data. However, it is at the expense of large time required for training. The optimization problem is in general non-convex and non-linear. However, smart techniques have been used to reduce the training time. For example, in pre-trained models, weights of initial layers are determined by using general classes of images like animals and birds. Using these weights, more specialized images could be classified by training only the final layers. For instance, pre-trained models could be used to further separate images of animals into dogs and cats.

9.9 Summary

Key ideas discussed in this chapter are summarized as follows:

- A binary linear classifier groups data points into two classes. The points belonging to these classes are separated by a hyperplane whose equation is written as $\langle \mathbf{w}, \mathbf{x} \rangle + \mathrm{b} = 0$. The weight vector **w** is normal to the hyperplane.
- The weight vector is computed using training data such that all the positive points satisfy $\langle w, x \rangle + b > 0$; negative points satisfy $\langle w, x \rangle + b < 0$.
- Prediction errors in unseen test points can be minimized by maximizing the margin. In maximal margin classification, the goal is to find a hyperplane which has the maximum distance from the nearest training data point.
- Maximal margin classification is solved as an optimization problem in which the norm of the weight vector is minimized, subject to the constraints that no training data points are misclassified.
- When the decision boundary separating points into different classes is not linear, it becomes a non-linear classification problem. Learning in the feature space is a popular technique for non-linear classification. Features are derived quantities computed by using the original input variables in linear or non-linear functions.
- Kernels are functions that help in non-linear classification by implicitly defining the feature vector. A kernel computes the dot product of two feature vectors. However, it is not always necessary to explicitly compute the feature vectors.
- Complex non-linear kernels involve tunable hyperparameters, and the results depend on the values of these model parameters. Complex kernels might overfit data and reduce generalizability.
- Support vector machines are a group of algorithms that permit learning in the feature space using kernels. The learning task is formulated as a quadratic optimization problem. The solution to the problem is a linear combination of a few input data points known as support vectors.
- Soft margin classification is used when the data are not linearly separable.

References

Cristianini, N. and Shawe-Taylor, J. (2000). *An Introduction to Support Vector Machines and Other Kernel-Based Learning Methods*, Cambridge: Cambridge University Press. https://doi.org/10.1017/CBO9780511801389

10 Inductive learning – decision trees and random forests

10.1 Introduction

Inductive learning attempts to discover rules from data. For example, if all the birds in a database have wings, an algorithm might conclude that "birds have wings". Such inferences tend to be temporary and might be reversed when more data are collected. In general, rules might be influenced by the bias in the data set. This is obvious from our everyday lives. We make generalized statements about people belonging to different races, speaking different languages, etc. Machines are also not immune to such biases.

The most popular use of inductive learning is for the classification task (Chapter 9). Rules to predict the value of the output variable might be extracted from data. A famous example of inductive learning is the ID3 algorithm in which a hierarchy of rules, known as a decision tree, is generated. Over the years, many variations of this algorithm have been developed, and these are widely used for practical engineering tasks. C4.5 algorithm is an extension of ID3.

10.2 ID3 algorithm

The ID3 algorithm recursively separates data into groups such that within each group, data are homogeneous. A group is homogenous if all the elements of the group have the same value for the output variable. The algorithm looks for discriminating features, that is, input variables that separate data into groups that are homogenous.

Consider the data in Table 10.1. The input variables are span and shape. The output variable is the slab system. Which input variable separates the output into homogenous sets? If span is chosen as the variable to divide the data into subgroups, the group corresponding to long span is completely homogenous because all the elements of this group have waffle slabs. The group corresponding to short span has a mixture of two-way and one-way slabs. If shape is chosen as the variable for dividing the data into subsets, both square and rectangular shapes have non-homogenous subsets.

Table 10.1 Selection of a Slab System for a Room

Span	*Shape*	*Slab system*
Long	Square	Waffle
Long	Rectangle	Waffle
Short	Square	Two-way
Short	Rectangle	One-way

DOI: 10.1201/9781003165620-12

This brings out the requirement for a measure of homogeneity, a metric that can tell how homogenous a group is. ID3 uses the concept of entropy. Shannon's entropy function was developed to quantify the information content in a data set and is a measure of in-homogeneity in a collection. The entropy is 0 when the set is perfectly homogeneous and 1 when the set is perfectly inhomogeneous. The definition of entropy is as follows:

$$H = -\sum_i p_i \log_2 p_i \qquad \textit{Eq 10.1}$$

where, H is the entropy of the output variable and p_i is the probability of i-th value of the output variable. The summation is performed over all the values of the output variable, i. The probability is calculated using the formula,

$$p_i = \frac{N_i}{N} \qquad \textit{Eq 10.2}$$

where N_i is the number of data points in the set having the i-th value of the output variable, N is the total number of points in the set. If all the points have the same value, then the probability is equal to one. Since the logarithm of one is zero, the entropy becomes zero. If there are two possible values for the output variable and the points are equally distributed, the probability of each value is (1/2), and the resulting entropy will have the maximum value of 1. The plot of entropy for the binary output case is given in Figure 10.1.

Shannon's entropy function is an important concept in information theory and is widely adopted for various tasks in computer science. For example, it has been used to determine the best location of sensors in monitoring tasks (Soman et al., 2017; Papadopoulou et al., 2014). In the ID3 algorithm, it is used to determine which input variable results in subsets having maximum homogeneity. The algorithm is summarized as follows:

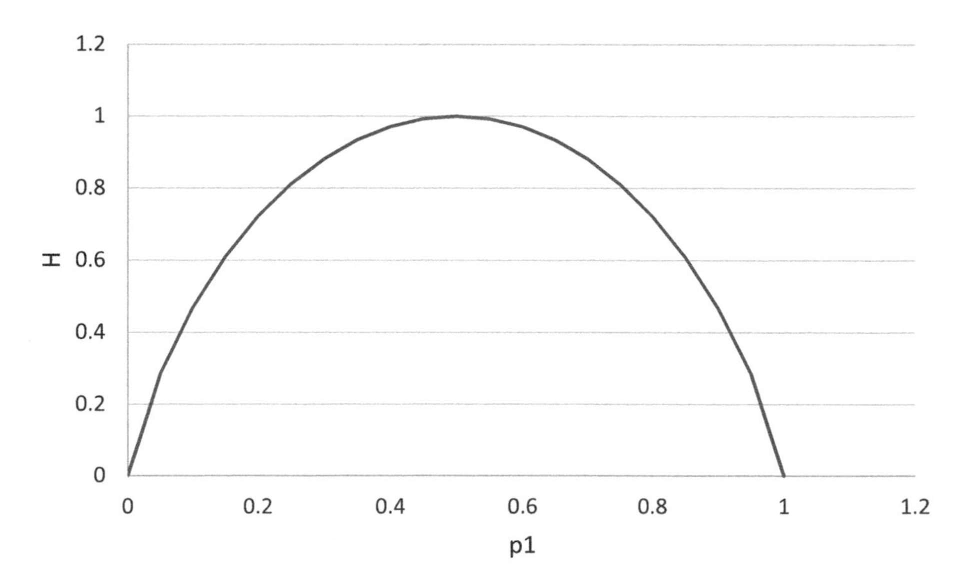

Figure 10.1 The entropy for the binary output case

ID3 Algorithm Part 1

Step 1: Create the root node of the decision tree containing all the data points. Set this as the current node to be subdivided.
Step 2: Call the recursive algorithm *DivideNode* (as follows) for the current node.

Algorithm *DivideNode*

Step 1: If the current node is homogenous, return without any further division.
Step 2: Calculate the initial entropy of the data in the current node. Assign this value to the best entropy.
Step 3: Loop 1: Repeat for each input variable i

Step 4: Loop 2: Repeat for each value j of the input variable i

Step 5: Create the subset of the data points in the current node having the value j of variable i. Compute the entropy of the subset. If the entropy is lower than the best entropy, assign best entropy to this value, select the value j of variable i as the best variable-value pair for subdivision.

End of Loop 2

End of Loop 1

Step 4: Using the best variable-value pair, divide the data in the current node into subsets. Add children nodes to the current node corresponding to these subsets. Recursively call *DivideNode* algorithm for each child node.

Exercise 10.1

Using Table 10.1, calculate the entropy if

a Span is chosen as the attribute
b Shape is chosen as the attribute

Explain which variable is chosen for dividing the data into homogenous subsets.

Answer

a If span is chosen as the attribute:

There are two possible values for span, long and short. The long span has entropy 0, since the output value is always waffle. The short span has entropy 1.0 since the output values are equally distributed between one-way and two-way. The average entropy for the variable span is $(0 + 1)/2 = 0.5$

b If shape is chosen as the attribute

Again, there are two possible values. The value square results in a subset having equal distribution of waffle and two-way slabs. Its entropy is 1.0 The value rectangle also results in entropy 1.0. Therefore, the average entropy is 1.0.

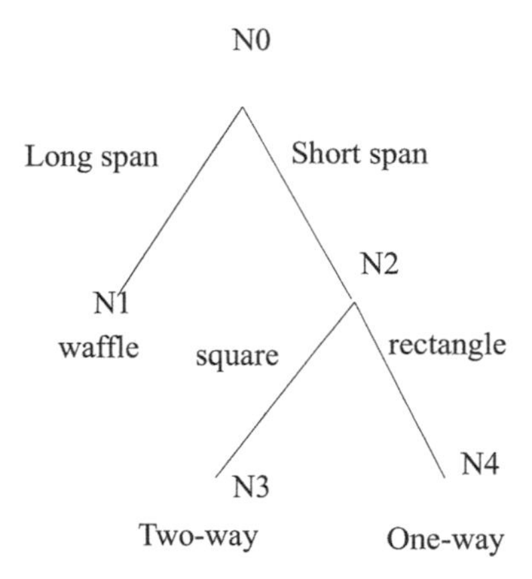

Figure 10.2 The decision tree for selecting a slab system for a room

Hence, dividing the set using the variable span results in lower entropy. The final decision tree is shown in Figure 10.2.

Exercise 10.2

Discuss whether variance or standard deviation can be used as a measure of homogeneity.

Exercise 10.3 Selecting a motor for a robot

Table 10.2 Data for Exercise 10.2

	Torque (Nm)	*Torque Class*	*Continuous Rotation*
Stepper	0.065	Low	No
Stepper	0.9	Low	No
DC Servo	6.8	High	Yes
AC Servo	1.27	Low	Yes

Form a decision tree to select the type of motor for a robot using the data in Table 10.2.

10.2.1 The discriminant

In Chapter 9, we saw that the discriminant in the case of a maximal margin classification is a hyperplane. That is, a linear combination of all the input variables determines the function that separates the points into classes. In the case of a decision tree, the values of a single variable determine the separation at each level within the tree. Figure 10.3 shows how the discriminant looks like in two dimensions. In general, the ID3 algorithm separates the solution space into boxes such that each box contains points that are as homogenous as possible.

Exercise 10.4

Study the data in Table 10.3 related to the accident track record of workers in a project:

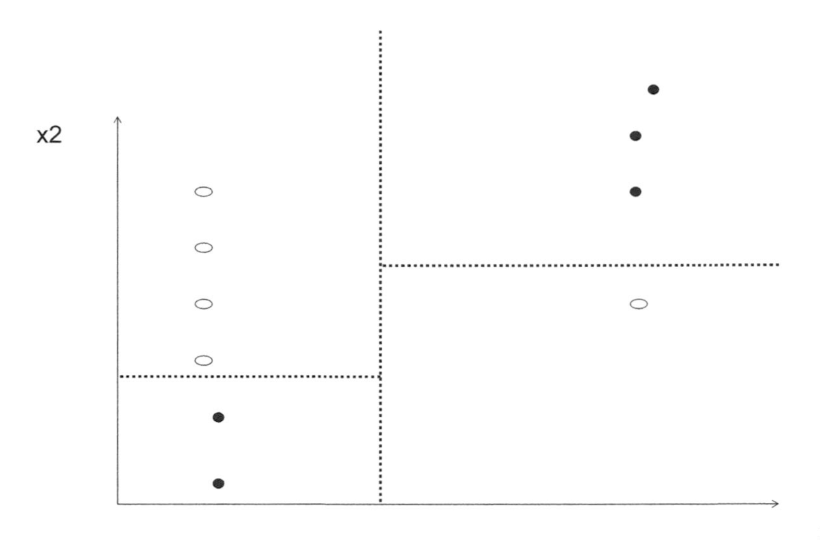

Figure 10.3 The discriminant function for ID3 algorithm

Table 10.3 Data for Exercise 10.4

Worker	*Age (years)*	*Experience (years)*	*Accidents*
A	20	1	1
B	40	15	0
C	25	3	1
D	29	4	1
E	28	5	1
F	42	1	1
G	32	5	0
H	35	10	0

We want to predict whether a worker is likely to cause accidents on site. Create a decision tree using the ID3 algorithm. Comment about the nature of the decision surface separating the two classes. Perform maximal margin classification for the same data. Compare the decisions surfaces of the two methods.

10.2.2 Continuous variables

The original ID3 algorithm was developed for discrete input variables. The decision tree algorithm has been extended for the case of continuous input variables. In this case, the value of the variable at which the set is subdivided should be determined. Greedy algorithms have been used to find the best location for division. A method to do this is to start with a random value for the separation; then, the value is iteratively modified such that there is improvement in the net entropy of the two subsets.

Decision trees can also be generated for continuous output variables. In this case, the range of values of the output variable is divided into multiple intervals, probability is computed for each interval, and the entropy is calculated.

10.2.3 Overfitting

Decision trees are susceptible to overfitting. In some cases, large number of rules are generated. This happens mainly when the decision boundary cannot accurately be represented as boxes. When the actual decision boundary is a hyperplane (whose equation depends on several variables), the decision tree algorithm tries to divide the data set using the values of individual variables. Hence many rules are generated, and the final data sets contain very few data points. In the worst case, the leaf nodes in the decision tree contain only one data point each. This is equivalent to creating one rule for each data point. Predictions made using only a few data points are likely to be unreliable. This could result in overfitting. While the decision tree predicts correctly when the points in the training data are presented for prediction, it is unable to make predictions when slight variations of these points are presented.

10.3 Random forests

Decision trees are sensitive to the values of variables in individual data points. The structure of the tree could change completely if a single data point is removed from the data set, especially if this data point influences the entropy of the subsets at the top nodes of the decision tree. In this case, all the subsequent divisions will change because the data sets for all the remaining nodes will be different.

The concept of *ensemble* methods has been developed to make use of multiple machine learning models to improve the accuracy of predictions. This avoids the problem of a single model getting influenced by specific data points. By using multiple models, errors in individual models might get canceled and more accurate prediction is obtained. Each individual model might be a *weak learner* in the sense that they are not capable of making accurate predictions because of the biases in the underlying model. By combining several weak learners, we create a "strong learner" that does not suffer from these biases. This results in more robust prediction.

The accuracy of a decision tree can be improved by taking the average prediction from many decision trees. Multiple trees make a "forest". Several trees can be generated randomly by using different subsets of data. Hence the term *random forest* is used. Random forests can potentially overcome the problem of overfitting because predictions are not influenced by the idiosyncrasies of individual data points.

Variations in the structure of the trees in a forest are increased using the following techniques: First, each tree is created using a different data set. Different data sets are generated from the same initial data using the technique of *bootstrapping*. Bootstrapping is the process of randomly sampling the original data with replacement. That is, the same data point might be selected more than once in the process. Second, the deterministic method of selecting the variable to divide the data set in the ID3 algorithm is replaced by a random method. Instead of selecting the variable that gives the minimum entropy, variables are randomly selected.

When a new test point is presented for prediction, all the trees are provided with the opportunity to make a prediction without getting influenced by each other. Just as the individuals in a committee make decisions individually without consulting with each other and cast their vote, the predictions of individual trees are computed in parallel. All the predictions are aggregated to compute the final prediction, which is a single number. The technique of using multiple models to make predictions in parallel and aggregating the results is known as *bagging*. This term should be distinguished from the term *boosting*. In boosting, multiple machine learning models are executed one after the other, to improve the accuracy of individual models. The two concepts are illustrated in Figure 10.4.

Exercise 10.5

To illustrate the process of using multiple machine learning models to make a prediction, consider the data in Table 10.1. This is a small data set. Not much variation in data is possible through bootstrapping. Consider three sets of data that are obtained by sampling the original data with replacement (Table 10.2). The same data point might be sampled more than once.

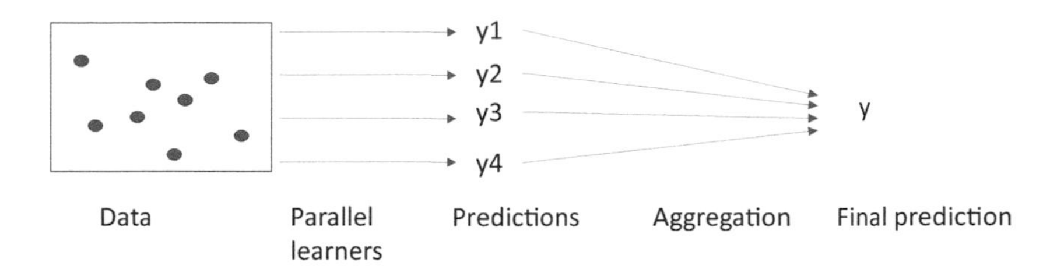

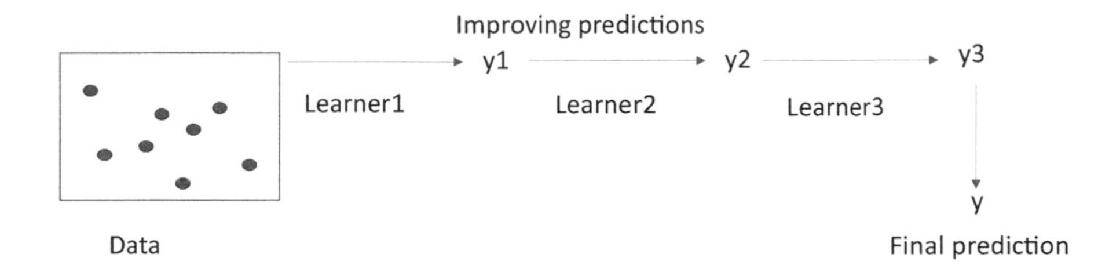

Figure 10.4 Bagging and boosting

Table 10.4 Selection of a Slab System for a Room

Data set 1

Span	*Shape*	*Slab system*
Long	Square	Waffle
Long	Rectangle	Waffle
Short	Square	Two-way
Short	Square	Two-way

Data set 2

Span	*Shape*	*Slab system*
Long	Square	Waffle
Long	Rectangle	Waffle
Short	Rectangle	One-way
Short	Rectangle	One-way

Data set 3

Span	*Shape*	*Slab system*
Long	Square	Waffle
Long	Rectangle	Waffle
Short	Square	Two-way
Short	Rectangle	One-way

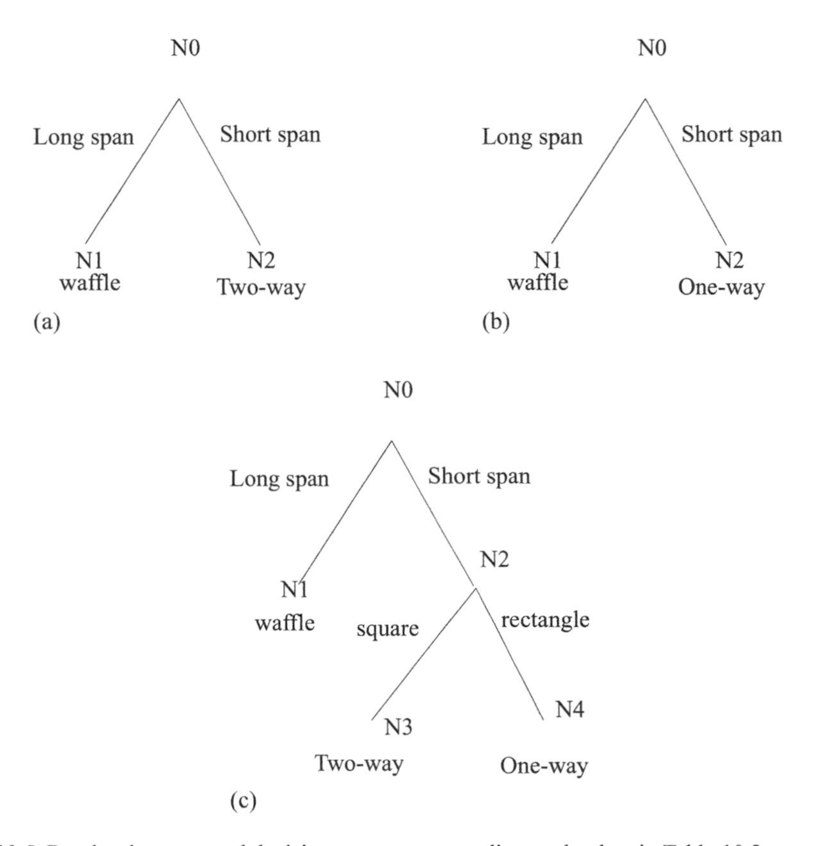

Figure 10.5 Randomly generated decision trees corresponding to the data in Table 10.2

By randomly selecting the variables for dividing each data set, three decision trees are obtained as shown in Figure 10.5.

For a new project, we need to select a slab system for a room which has short span and square shape. According to the first tree, the prediction will be two-way, since the shape does not appear in the decision-making. On the contrary, the prediction will be one-way using the second decision tree. The third decision tree uses the original data and predicts the two-way slab system. Taking the prediction with the highest vote, the resulting prediction of the random forest will be the two-way slab system. Even though, the second tree made a mistake in the prediction, the "wisdom of the crowd" ensured that the prediction of the ensemble is correct.

10.4 Summary

Inductive learning is the process of discovering generalized rules from data. ID3 is one of the oldest inductive learning algorithms for generating decision trees. More sophisticated algorithms such as random forests were developed later using the concept of ensemble learning. Ensemble learning makes use of multiple machine learning models that are executed in parallel (bagging) or in sequence (boosting). Random forests use bagging to generate multiple predictions from several decision trees, and these are aggregated to compute the final prediction. Biases in individual models are eliminated in this process and more robust performance is achieved.

References

Papadopoulou, Maria, Raphael, Benny, Smith, Ian F.C., and Sekhar, Chandra. (2014). Hierarchical Sensor Placement Using Joint Entropy and the Effect of Modeling Error, *Entropy*, 16, pp. 5078–5101.

Soman, Ranjith K., Raphael, Benny, and Varghese, Koshy. (2017, March). A System Identification Methodology to Monitor Construction Activities Using Structural Responses, *Automation in Construction*, 75, pp. 79–90.

11 Unsupervised learning algorithms

11.1 Introduction

Several unsupervised machine learning algorithms were qualitatively introduced in Chapter 6. These techniques are mainly used to generate knowledge about relationships and patterns in data. For example, it is useful to find out common factors in a data set; factor analysis is routinely performed by researchers (Lim et al., 2012). This chapter discusses the mathematical details of some commonly used unsupervised learning algorithms.

11.2 Principal Component Analysis (PCA)

PCA was introduced in Section 6.2.2.3. The goal of principal component analysis is to identify the most important components that can explain most of the variance in the data. If data points are strongly aligned in a certain direction u1, the variance will be maximum in this direction; the variance in directions perpendicular to u1 will be low. PCA helps to discover the directions in which there is maximum variance, that is, the principal directions in which data are strongly aligned. Through PCA, we get a new set of variables (principal components) that are linear combinations of the initial variables. The principal components are the eigenvectors of the data covariance matrix. The eigenvalues are the variances in these directions. If the eigenvalues are ordered in the descending order, the corresponding eigenvectors represent directions in which there is maximum variance in data.

Note: In the following discussion, it is assumed that the data are centered, that is, the mean is zero. This can be done by subtracting the mean of each variable from the value of the variable, that is, by shifting the origin to the mean value. Then the covariance matrix is written as

$$COV = \frac{X^T X}{m-1} \quad \textit{Eq 11.1}$$

where m is the number of data points. Using our usual notation, the data matrix X is of dimensions (m x n), where n is the number of variables.

If λ_1, λ_2 etc. are the eigenvalues of the covariance matrix sorted in the descending order, and $v_1, v_2, \ldots, v_n$ etc. are the corresponding eigenvectors which are normalized (having unit length), the eigenvectors can be arranged in the form of a square matrix V as follows:

$$V = [\nu_1 \nu_2 \ldots \nu_n] = \begin{bmatrix} \nu_{11} & \cdots & \nu_{n1} \\ \vdots & \ddots & \vdots \\ \nu_{1n} & \cdots & \nu_{nn} \end{bmatrix} \quad \textit{Eq 11.2}$$

DOI: 10.1201/9781003165620-13

Where v_{n1}, v_{n2} etc. are the components of the n-th eigenvector v_n.

Multiplying the covariance matrix with each eigenvector v_i gives the same vector scaled by the corresponding eigenvalue λ_i, Individual equations for each eigenvector can be combined into a matrix equation as follows:

$$\frac{(X^T X)}{m-1} V = \frac{(X^T X)}{m-1} [v_1 \quad v_2 \quad \dots \quad v_n]$$

$$= [v_1 \quad v_2 \quad \dots \quad v_n] \begin{bmatrix} \lambda_1 & 0 & \dots & 0 \\ 0 & \lambda_2 & & 0 \\ & & \dots & \\ 0 & 0 & & \lambda_n \end{bmatrix}$$

Therefore,

$$[COV]V = \frac{(X^T X)}{m-1} V = VL \qquad Eq\ 11.3$$

Where L is a diagonal matrix containing the eigenvalues as the diagonal elements, and all the non-diagonal terms are zero.

A property of eigenvectors is that they are orthogonal. That is, the dot product of two eigenvectors is zero, unless they are the same vector. Therefore, the multiplication of the V matrix with its transpose V^T on theix with its transpose V^T on ix with its transpose V^T on containing 1 along the diagonals and 0 elsewhere. That is,

$$V^T V = \begin{matrix} 1 & 0 & \dots & 0 \\ 0 & 1 & & 0 \\ & & \dots & \\ 0 & 0 & & 1 \end{matrix} \qquad Eq\ 11.4$$

The eigenvectors represent an orthogonal coordinate system which is rotated with respect to the original coordinate system. The coordinate axes are rotated such that the first axis has the maximum variance in data, the next axis has the next highest variance, and so on. The coordinates of the data points in this rotated coordinate system can be calculated by taking the dot product of the data points with the unit vectors in each rotated direction. This is equivalent to multiplying data matrix X with the V matrix as follows:

$$U = X V \qquad Eq\ 11.5$$

Writing X and V in terms of their components,

$$U = X V = \begin{bmatrix} x_{11} & \cdots & x_{n1} \\ \vdots & \ddots & \vdots \\ x_{1n} & \cdots & x_{nn} \end{bmatrix} \begin{bmatrix} v_{11} & \cdots & v_{n1} \\ \vdots & \ddots & \vdots \\ v_{1n} & \cdots & v_{nn} \end{bmatrix} \qquad Eq\ 11.6$$

The i-th row in the X matrix represents the i-th data point. The columns represent the variables x1, x2, . . . , xn. Multiplying a row *i* of the X matrix with a column *j* of the V matrix is equivalent to taking the projection of the vector representing the data point *i* on the eigenvector *j*. This gives the j-th coordinate of the data point *i* in the rotated coordinate system.

Post-multiplying the inverse of V, V^{-1} on both sides of the equation in Eq 11.5,

$$U V^{-1} = X V V^{-1} \quad \text{Eq 11.7}$$

Since $V V^{-1}$ is equal to the identity matrix, this equation becomes

$$U V^{-1} = X \quad \text{Eq 11.8}$$

Since the inverse of the matrix V is its transpose from Eq 11.4,

$$X = U V^T \quad \text{Eq 11.9}$$

That is, the original coordinates X are retrieved by multiplying the new coordinates U with the transpose of the V matrix.

Let us calculate the covariance matrix in the rotated coordinate system,

$$COV_U = \frac{U^T U}{m-1} = \frac{(X V)^T (X V)}{m-1} = \frac{\left(V^T X^T\right) X V}{m-1} = V^T \frac{X^T X}{m-1} V = V^T \left[COV\right] V \quad \text{Eq 11.10}$$

That is the covariance matrix in the rotated coordinate system is obtained by transforming the original covariance matrix by multiplying by V and V^T. Using Eq 11.3.

$$COV_U = V^T \left[COV\right] V = V^T (V L)$$

Since the eigenvectors are orthogonal, this expression becomes,

$$COV_U = L = \begin{bmatrix} \lambda_1 & 0 & \dots & 0 \\ 0 & \lambda_2 & & 0 \\ & & \dots & \\ 0 & 0 & & \lambda_n \end{bmatrix} \quad \text{Eq 11.11}$$

That is, the covariance matrix in the rotated coordinate system is a diagonal matrix containing the eigenvalues. Since the non-diagonal terms are zero, the variables in the rotated coordinate system are un-correlated. The diagonal terms represent the variance in the rotated coordinates. Hence, the eigenvalues are the variances of the principal components.

What is the expression for the covariance of the data matrix in the original coordinate system, in terms of the new coordinates?

$$COV = \frac{X^T X}{m-1} = \frac{\left(UV^T\right)^T UV^T}{m-1} = \frac{\left(VU^T\right)UV^T}{m-1} = \frac{V(U^T U)V^T}{m-1} = V\,L\,V^T \quad \text{Eq 11.12}$$

After finding the eigenvalues and the eigenvectors, the covariance matrix can be approximated as follows:

$$COV \approx \lambda_1 v_1 v_1^T + \lambda_2 v_2 v_2^T + \cdots = \sum_{i=1}^{N} \lambda_i v_i v_i^T \quad \text{Eq 11.13}$$

The significance of Eq 11.13 is that the covariance matrix in the original coordinate system can be written as a linear combination of a series of square matrices that have the eigenvalues as the coefficients. The last eigenvalues are small, contributing very little to the covariance matrix; they can be omitted. The covariance matrix can be reasonably approximated using only the first few eigenvalues and eigenvectors. This helps in data compression because several dimensions (corresponding to small eigenvalues) might be eliminated in the analysis, and the size of data becomes small.

Example

Consider the example discussed in Section 6.4.3.2. The plot of the two variables in the problem is shown in Figure 11.1. The direction u1 is obtained as the eigenvalue of the covariance matrix of this data. This is the direction along which there is maximum variance in the data as seen by the spread of the points in this direction. The eigenvalue of this eigenvector has

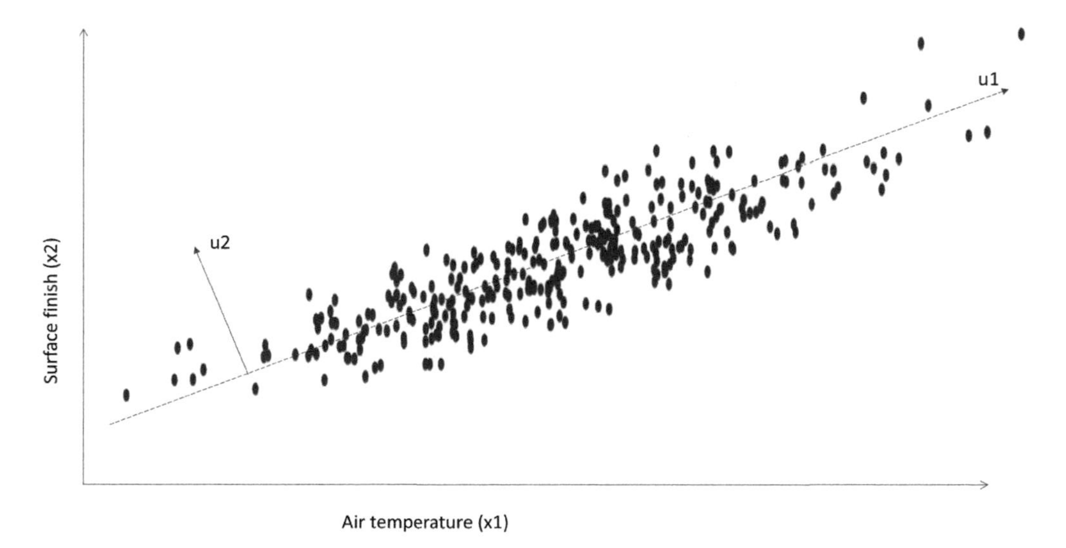

Figure 11.1 Rotated coordinate system

the maximum value, which is nothing but the variance in this direction. The perpendicular direction u2 is the next eigenvector. The spread of data in this direction is the minimum and consequently, the corresponding eigenvalue will be small. If all the points are strongly aligned in the u1 direction with no spread in the u2 direction, the second eigenvalue will be zero.

PCA helps in dimensionality reduction, visualization, compressing data, and for identifying most important features in a data set (Saitta et al., 2005). Some of these possibilities are illustrated in the examples.

Exercise 11.1

Consider the data in Table 11.1. There are three variables x1, x2, and x3. We want to verify whether there is multi-collinearity in the data and in that case, find new variables that are not correlated. We will use PCA for this task.

Table 11.1 Data for Exercise 11.1

x1	*x2*	*x3*
0.98	2.01	1.49
0.48	1.01	0.77
2.01	3.99	3.02
0.59	1.2	0.88

The principal components are the eigenvectors of the covariance matrix. So, we need to first compute the covariance matrix. From Eq 11.1, the covariance matrix is:

$$COV = \begin{matrix} 1.4581 & 2.84935 & 2.1668 \\ 2.84935 & 5.569275 & 4.235 \\ 2.1668 & 4.235 & 3.2214 \end{matrix} \qquad \textit{Eq 11.14}$$

Computing the determinant of the matrix, we get 1.74e−6, that is, close to zero. It shows that there is collinearity in the data. PCA is used to determine the directions in which there is no variability in data and to create new variables that are not correlated.

We need to compute the eigenvalues of the covariance matrix. The general equation is given in Section 7.3. However, it is difficult to solve a cubic polynomial to compute the eigenvalue. Instead, a numerical method is introduced here. It is called the power method. It makes use of the feature that when a matrix C is multiplied by its eigenvector, you get the same vector multiplied by a constant, which is the corresponding eigenvalue. Any arbitrary unit vector z can be written as a linear combination of the eigenvectors as follows:

$$z = a_1 v_1 + a_2 v_2 + \cdots \qquad \textit{Eq 11.15}$$

If an arbitrary unit vector is multiplied by the matrix, the coefficients in the linear combination get multiplied by the corresponding eigenvalues. That is,

$$\begin{aligned} C\,z &= C[a_1 v_1 + a_2 v_2 + \cdots] = [a_1 C v_1 + a_2 C v_2 + \cdots] \\ &= [a_1 \lambda_1 v_1 + a_2 \lambda_2 v_2 + \cdots] \end{aligned} \qquad \textit{Eq 11.16}$$

If this multiplication is performed repeatedly, the coefficient of the principal eigenvector gets amplified significantly compared to the remaining eigenvectors, because of the relatively large magnitude of its eigenvalue. After two multiplications, the expression becomes:

$$C\,(Cz) = C[a_1\lambda_1 v_1 + a_2\lambda_2 v_2 + \cdots] = [a_1\lambda^2{}_1 v_1 + a_2\lambda^2{}_2 v_2 + \cdots] \qquad \textit{Eq 11.17}$$

Here, the coefficient of v_1 has got magnified by $\lambda^2{}_1$. The remaining terms become small compared to the first term. Therefore, repeated multiplication of the matrix with an arbitrary unit vector has the effect of rotating this vector in the direction of the principal eigenvector. This principal is used in the power method to extract the eigenvector and the corresponding eigenvalue. This is illustrated in the following.

Take an arbitrary vector as follows:

1
1
1

Normalize this by dividing by the length of this vector to get a unit vector as follows:

0.57735
0.57735
0.57735 *Eq 11.18*

Multiplying this unit vector with the matrix COV in Eq 11.14,

3.73791
7.305574
5.555957 *Eq 11.19*

The length of this vector is 9.91. Dividing by the length of the vector, the new rotated unit vector is obtained as:

0.377178
0.737177
0.56063

If this vector is compared with the initial vector we started with, it is noticed that the vector has been rotated because the initial guess of the unit vector is very different from the principal eigenvector. Repeating the procedure, multiplying the covariance matrix by the previous unit vector and normalizing, we get,

0.377173
0.737182
0.560628

If the process is repeated, it is noted that this unit vector does not get rotated significantly and the final eigenvector *v1* is obtained as:

0.377173
0.737182
0.560628 *Eq 11.20*

When this vector is multiplied by the covariance matrix, the vector simply gets scaled by a constant, that is, the length of the vector changes without any change in the direction. The scaling factor is the eigenvalue corresponding to this vector. Since we started with a unit vector, the scaling factor is just the length of the resulting vector after the multiplication of the covariance matrix. The principal eigenvalue thus obtained is 10.248.

The original data can be transformed into a rotated coordinate system in which the first coordinate axis is along the first eigenvector, and so on. We can reasonably approximate the original data by using only the first axis and ignoring the remaining directions in which the data has low variance. That is, we assume that all the data points lie on the axis representing the first eigenvector, and small deviation of the points in the perpendicular direction is ignored. We can see that there is no significant loss of information when you do this. If more accuracy is needed, more eigenvectors might be used.

We can calculate the coordinates of all the data points in the rotated coordinate system in which the principal axes correspond to the eigenvectors that have been obtained. The new coordinates can be computed in this coordinate system by taking the projection of each data point on the new axes, that is, by taking the dot product of each data point and each eigenvector. Multiplying the data in Table 11.1 with the eigenvector shown in Eq 11.20, the projections of these points in the direction of the first principal component are obtained as:

$$\begin{matrix} 2.6867 \\ 1.35728 \\ 5.392569 \\ 1.600503 \end{matrix} \qquad \text{Eq 11.21}$$

These are the coordinates in the rotated coordinate system (using only the first axis and ignoring the remaining axes in the perpendicular directions).

To check how much loss in accuracy is caused by ignoring the second and third eigenvectors, we can transform the coordinates back into the original coordinate system. Using Eq 11.9, the approximate values of variables in the original coordinates are obtained as:

$$\begin{matrix} 1.013351293 & 1.980586 & 1.506239 \\ 0.511929622 & 1.000562 & 0.760929 \\ 2.033932365 & 3.975303 & 3.023225 \\ 0.603666671 & 1.179861 & 0.897287 \end{matrix}$$

Computing the difference between this matrix and the original data in Table 11.1, the error in approximating the data using a single principal component is obtained as

$$\begin{matrix} 0.033351 & -0.02941 & 0.016239 \\ 0.03193 & -0.00944 & -0.00907 \\ 0.023932 & -0.0147 & 0.003225 \\ 0.013667 & -0.02014 & 0.017287 \end{matrix}$$

The errors are in the second decimal place, which might be reasonable in many applications. Therefore, the original data consisting of 4 rows and 3 columns can be compressed into a table consisting of 4 rows and 1 column without much loss of accuracy. Instead of visualizing data in 3 dimensions, we are able to visualize data in a single dimension. If an output variable that depends on the input data in Table 11.1 needs to be visualized, we can make a 2D scatter plot in which the x-axis is the first principal component values (Eq 11.20) and the y-axis is the output variable. The plot will represent the projected view of data on a plane containing the eigenvector, that is looking in a direction perpendicular to the principal component.

To calculate the next eigenvalue, subtract the component from the first eigenvalue from the covariance matrix using Eq 11.13. The resultant matrix does not contain the first eigenvalue we computed earlier. Hence, by repeating the process using the resultant matrix, the process will converge to the second eigenvector.

First compute $v1\ v1^T$ using Eq 11.20:

0.142259601	0.278045	0.211454
0.278045122	0.543437	0.413285
0.211453818	0.413285	0.314304

Subtracting $\lambda_1 v_1 v_1^T$ from the original covariance matrix, the new matrix is obtained as:

0.000245274	−1.4e−05	−0.00015
−1.406e−05	0.000218	−0.00028
−0.00014652	−0.00028	0.000464

Finding out the eigenvalue of this matrix by repeating the same process as before, we see that the eigenvalue converges to 0.000676. This means that there is not much variation in the direction perpendicular to the first eigenvector. There are small variations in this direction, due to which the determinant of the covariance matrix is not exactly zero, but a small value. This causes numerical instabilities while applying methods like ordinary least squares regression. A robust method of generating a model in such cases is principal component regression, which is discussed later. The second eigenvector is obtained as:

−0.25963
−0.4969
0.828058

Application: principal components in image data

If images are created by translation or rotation of a particular shape, the pixel values change at the locations where the shape has moved. Other pixels will have the same value. Hence, all the images can be expressed as small variations of a base image.

The first principal component contains non-zero values for pixels that are different from the base image. It is an image which, when added to the base image, represents an average image representing common variations in the data set. Each image in the data set can be considered as a linear combination of some base images (whose pixel values form the initial eigenvectors).

11.2.1 Example: PCA of image data

Sixteen gray scale images containing 5 × 5 pixels are shown in Figure 11.1. The background color has a pixel value of 0, which is not shown in the figure. Only pixels that have non-zero values are shown. All the images contain a diagonal line having a different brightness value.

Each data point in this example consists of 25 values, the brightness of each pixel. The input vector consists of 25 variables. However, all the images are similar. PCA can be used to find out the common feature in all the images. The data matrix X consists of 16 rows (for 16 images) and 25 columns (for 25 pixels). Using this data, the covariance matrix is computed as follows:

Var	1	...	7	8	...	13	...	19	...	24	25
1	0	0									
...	0	0									
7			953			953			953		
8											
...											
13			953			953			953		
...											
19											
20			953			953			953		
...											
24										0	0
25										0	0

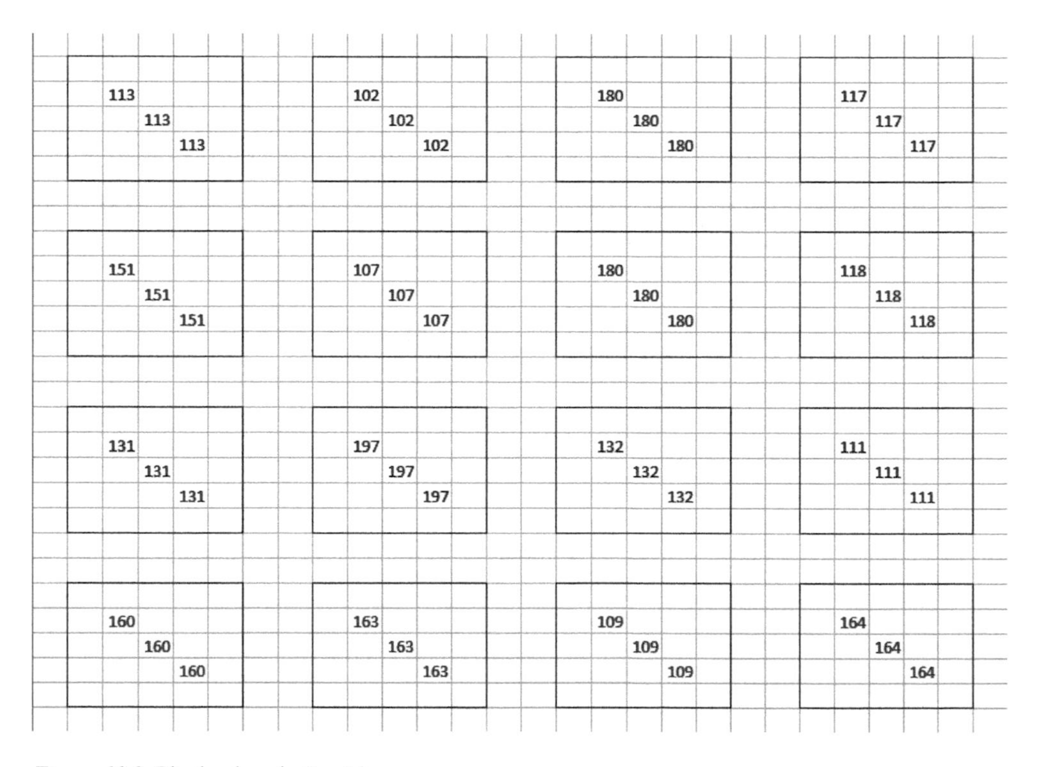

Figure 11.2 Pixel values in 5 × 5 images

All the components of the covariance matrix are zero except for the variables 7, 13, and 20, which correspond to the location of pixels that have non-zero values. The first eigenvector is the following:

#	Coefficient
1	0
2	0
...	0
7	0.57735
...	0
13	0.57735
...	0
19	0.57735
...	0
24	0
25	0

All the coefficients of the first eigenvector are zero except at the positions 7, 13, and 19, which correspond to the non-zero pixels in the image. If the coefficients of the eigenvector are arranged in the form of a 5 × 5 matrix, it looks like this:

		0.57735				
			0.57735			
				0.57735		

It resembles the images in Figure 11.1. Hence the first principal component represents an image that shows the variation with respect to a base image having the average pixel values. All the images in Figure 11.1. are obtained by multiplying the first principal component by a specific constant, which is the coordinate of the point along the direction represented by the eigenvector. Therefore, instead of storing the coordinates of all the pixels in each image, a single number needs to be stored, and the entire image can be reproduced. This can be used to compress the image data.

11.2.2 Example: PCA of text data

How can PCA be used to compress the following text

fool fool fool fool fool
fool fool fool fool fool
tool fool fool fool fool

fool fool fool fool fool
fool fool fool fool fool

We can create a "bag of words" from the document. The document contains two words: fool and tool. Let us represent these words using the indices 0 and 1. Using this representation, each line in the document can be written as a vector whose components are either 0 or 1 depending on the word. Using this data matrix, the eigenvectors are computed. Similar to the previous example, all the data points (lines in the document) can be written as a linear combination of the principal components.

11.2.3 Principal component regression

When data has multi-collinearity, ordinary least squares regression does not work correctly. In such cases, regression might be carried out using the principal components. That is, instead of using the original variables for regression, the principal components are used as secondary variables (features). Since principal components are orthogonal to each other, collinearity in data are avoided. In addition, only a few principal components are used, thereby reducing the dimensionality of the problem. This procedure is also more robust because the directions in which there are small variations due to noise are ignored.

Example

Consider the following data with input variables x1, x2 and x3. y is the output variable:

x1	x2	x3	y
−0.76	−1.42	2.18	14
−0.56	−0.42	0.98	15.4
0.24	−0.22	−0.02	17.2
0.34	1.18	−1.52	18.8
0.74	0.88	−1.62	19.3

The data are centered; all the input variables have zero mean. We want to develop a linear model to predict the value of the output variable y. Notice that the data are collinear; the variable x3 is a linear combination of x1 and x2. But assume that we are not aware of this condition. Performing ordinary linear regression, the ($X^T X$) matrix is computed as follows, after adding a column consisting of ones at the end:

1.6120	2.3140	−3.9260	0.0000
2.3140	4.4080	−6.7220	0.0000
−3.9260	−6.7220	10.6480	0.0000
0.0000	0.0000	0.0000	5.0000

The determinant of this matrix is computed using a popular spreadsheet program to be 2.33293E−14. This is close to zero. It is not exactly zero because of roundoff errors in floating point calculations. Without knowing that the determinant is zero, when the inverse of the matrix is computed using the spreadsheet program, the inverse is obtained as follows:

3.7530E+14	3.7530E+14	3.7530E+14	−8.3333E−02
3.7530E+14	3.7530E+14	3.7530E+14	−8.3333E−02
3.7530E+14	3.7530E+14	3.7530E+14	−8.3333E−02
−8.3333E−02	−8.3333E−02	−8.3333E−02	2.0000E−01

The coefficients are large because of numerical errors. Multiplying the inverse of the matrix with the right-hand side of the regression equation (Eq 8.11), we get the weight factors

−0.05833
0.941667
−0.05833
16.94

The first three elements are the coefficients of x1, x2 and x3. The last element is the bias term b. The actual values of these coefficients are [3, 2, 1 and 16.94) which can be verified by multiplying these coefficients with the values of x1, x2 and x3 with the original data. Exact values are predicted using these coefficients. However, regression has resulted in a different set of coefficients because of numerical roundoff errors. Using the weight factors obtained through regression, the actual (y) and predicted values (yp) are shown in the following table:

y	yp
14	15.52
15.4	16.52
17.2	16.72
18.8	18.12
19.3	17.82

Large errors are observed in most data points.

Now let us select a subset of features using PCA. First the covariance matrix is computed as follows:

0.403	0.5785	−0.9815
0.5785	1.102	−1.6805
−0.9815	−1.6805	2.662

Using the power method, the first eigenvalue is obtained as 4.087, and the first eigenvector is obtained as:

0.2953
0.5116
−0.8069

The second eigenvalue is much smaller, 0.0803. That is, there is very little variation in the direction perpendicular to the first principal component. The second eigenvector is:

0.7612
−0.6364
−0.1248

Using these as the coordinate axes, the original coordinates are transformed into the rotated coordinate system as follows:

u1	u2
−2.7099	0.0530
−1.1710	−0.2813
−0.0255	0.3252
1.9305	−0.3024
1.9759	0.2055

Note that, the number of dimensions is decreased by one because we have ignored the third principal component. As before, a new column consisting of ones is added at the end, and regression is performed. The following weight factors are obtained:

1.1022
0.8861
16.9400

Here, the first two elements are the coefficients of u1 and u2; the last element is the bias term. Using these coefficients, the predicted and actual values are compared as follows:

u1	u2	Yp	y
−2.7099	0.0530	14	14
−1.1710	−0.2813	15.4	15.4
−0.0255	0.3252	17.2	17.2
1.9305	−0.3024	18.8	18.8
1.9759	0.2055	19.3	19.3

There is an exact match between the predicted (yp) and actual (y) values. By taking the principal components as derived features, the collinearity in the data are avoided and numerical errors are eliminated.

11.3 Clustering

Clustering was introduced in Section 6.2.2.1. Details of clustering algorithms are covered in this section. In general, clusters are identified using a measure of similarity (or distance). Points are added to clusters such that the intra-class similarity is high, and the inter-class similarity is low (Saitta et al., 2008). That is, points that are most similar belong to the same cluster, and dissimilar points are added to different clusters.

There are two types of clustering algorithms.

1 Hierarchical clustering
2 Non-hierarchical clustering

11.3.1 Hierarchical clustering

In hierarchical clustering, the clusters are generated and organized in the form of a tree structure. Each node of the tree represents a cluster. The cluster at the top of the tree contains all the data points. This is subdivided into the clusters at the next level and so on. At the bottom, the leaf nodes contain single data points. Required number of clusters can be obtained by appropriately pruning the tree. In Figure 11.3, the clusters in a data set of six points are shown. At the second level, there are two clusters containing four and two data points. The number of data points in these two clusters is not balanced because the points a, d, e, and f might be close to each other compared to the other points. The other two points, b and c, might be close to each other, forming another cluster. These clusters are further subdivided into smaller clusters based on their relative distances. If we are interested in only three clusters, we can prune the tree using an appropriate algorithm. For example, (a,e), (d,f) and (b,c) could be selected as the three clusters.

Bottom-up hierarchical clustering is an algorithm for creating a hierarchy of clusters. A simple algorithm for this is:

Step 1: Start by assigning each data point to its own cluster.
Step 2: Find the closest pair of clusters and merge them into a single cluster. This will create a new cluster at the next higher level in the tree.
Step 3: Repeat steps 1 and 2 until all items are clustered into a single cluster.

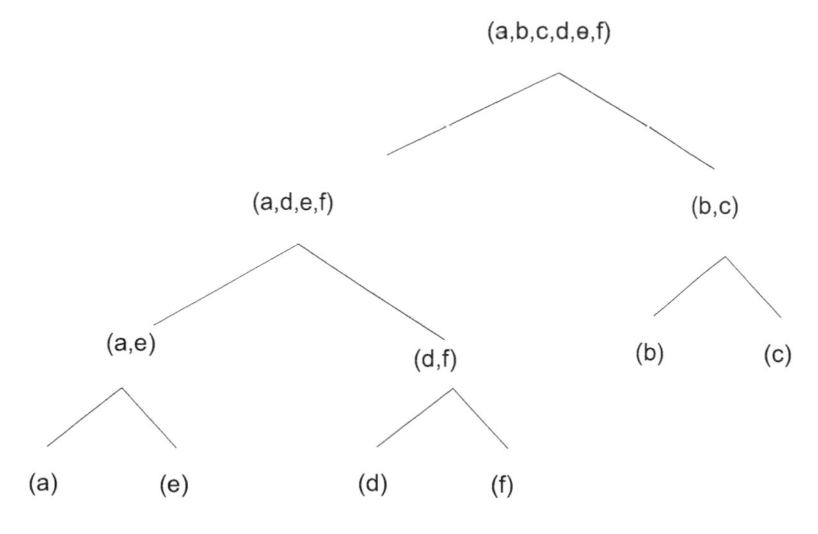

Figure 11.3 Hierarchical clustering

In Figure 11.3, there are six initial clusters containing one point each. These are a, b, c, d, e, f. Suppose a and e are the points that lie closest to each other. These two clusters are merged to create the node containing the set (a,e), which is the parent node of these two clusters in the tree. Similarly, d and f are merged to create the cluster (d,f) and so on. At the end of the process the node at the top contains all the data points.

11.3.2 Non-hierarchical clustering

In non-hierarchical clustering, all the data are grouped into a set of predefined number of clusters specified by the user. K-means, K-medians, Hard C-means, and fuzzy C-means algorithms are examples of this. Only K-means clustering is discussed here. In K-means clustering, clusters are created such that points that lie close by (according to some definition of distance) are grouped together. The required number of clusters, k, is usually specified by the user. The algorithm is summarized using the following steps:

Step 1: Randomly select k data points to be the initial centroids of clusters.
Step 2: Compute the distance of each point to the centroids of clusters. Assign each data point to the cluster whose centroid is the closest to it.
Step 3: Recompute the centroids of all the clusters based on the data points that have been assigned to them.
Step 4: Repeat steps 2 and 3 if the clusters have changed after the re-computation of the centroids.

Even though, the algorithm starts with random points for the centroids of the clusters, after a few iterations, the distances between the clusters increase, because the points that lie far away tend to pull the centroids towards them and the clusters tend to separate from each other.

11.4 Applications

Unsupervised learning has many applications in engineering, especially in automation applications. A few potential applications are mentioned here.

Patterns within images can be discovered using clustering. Groups of pixels that are similar are put into the same cluster. By examining the characteristics of the clusters, useful information can be extracted from images. For example, images of layers of 3D printed concrete elements can be analyzed using this technique. Variations in the thickness of layers and other changes in properties can be extracted.

An automated system for connecting steel components might make use of computer vision for locating bolt holes. Various unsupervised learning techniques could be used for tasks such as detecting cavities, edge detection, and locating the boundary of objects.

Data points might be clustered first, before performing regression. This helps to determine relationships that are specific to each cluster. For example, if an empirical relationship between the load and the deflection in a cracked concrete element is sought (See Figure 8.1), the data points might be first grouped together using clustering techniques. Then a separate regression is performed for each cluster. The clusters might be merged together, if the regression coefficients are similar.

Laser scanning results in large volumes of data consisting of point clouds. Many points might be similar and may not have significant variations. The data can be compressed using clustering and PCA. The analysis might be simplified using smaller data sets.

Simple techniques such as correlation might be used in analysis of video recordings. A video consists of many frames which are individual images. Moving objects can be identified in multiple frames through this method. The velocity of objects and other useful information can be extracted.

Many engineers and companies face the problem of information overload. There are too many documents containing valuable data. Locating the right piece of information from these documents is not easy. Searching using keywords results in many documents that should be manually examined for relevance. Clustering helps in grouping documents so that the processing becomes easier.

11.5 Summary

Unsupervised learning techniques such as PCA and clustering have been covered in detail in this chapter. Mathematical details and algorithms were explained using numerical examples. Potential applications in automation were briefly mentioned.

References

Lim, T.H.B, Ling, Y.Y.F., William, Ibbs C., Raphael, B., and Ofori, G. (2012). Mathematical Models for Predicting Organizational Flexibility of Construction Firms in Singapore, *Journal of Construction Engineering and Management-ASCE*, 138 (3), pp. 361–375.

Saitta, S., Raphael, B., and Smith, I.F.C. (2005). Data Mining Techniques for Improving the Reliability of System Identification, *Advanced Engineering Informatics*, 19 (4), pp. 289–298.

Saitta, S., Raphael, B., and Smith, I.F.C. (2008). A Comprehensive Validity Index for Clustering, *Intelligent Data Analysis*, 12 (6), pp. 529–548.

Epilogue

Ajay accepted a job offer at Intelligent Building Consultants, Chennai, primarily because he was impressed by the efforts taken by the company to improve the comfort and convenience of its employees, while caring for the environment. He was not disappointed; the job satisfaction he got was beyond anything he had expected. He worked on several interdisciplinary projects that involved coordination between civil, mechanical, and electrical engineers. His knowledge of automation and machine learning was immensely beneficial for proposing new solutions as well as ensuring their effective implementation. With his knowledge of fundamentals in computing as well as engineering, he could easily pick up emerging areas and experiment with new technologies. Soon he rose up the corporate ladder, and he is currently being considered for the position of chief technology officer of the company.

Index

3D printers 30, 33, 49, 50, 124, 125
6-DOF robot 49, 122

AC 18
actuators 12–15, 24–27, 31–33, 38–41, 48–60, 114–116
ADC 42, 52
AI 4, 5, 12, 137
air conditioning 83
alternating current 17, 18
ampere 17
analog input 39, 52, 53, 54
analog output 23, 24, 52, 54
analog to digital converter 42
anode 16
Arduino 21, 39, 40, 52–63
artificial intelligence 4, 5, 12, 137
autonomous cars 5

BACNET 39, 47, 48, 53–60
BAS 7, 8
baud rate 42, 44, 60, 61
BIM 12, 107
binary input 25, 52, 53, 54
binary output 23, 52, 54
Bluetooth 47, 105, 106
BMS 7, 8, 74, 80, 99, 100, 101, 102, 103
building automation 4, 7, 13, 51, 57, 58, 69
building information modeling 12, 107

cables 22, 23
cathode 16
CO_2 23
communication 7–15, 41–68, 106
communication networks 7, 15, 41, 55
controller 12–30, 38–47, 52–64

damper(s) 6, 30, 31, 64, 65, 89, 90, 101
DC 18, 19
DC motor 27, 28
DDC 38–39
digital input 25, 53
digital output 23, 53, 54
direct current 16
direct digital controller 38–39
drones 11, 115

electrical actuators 25
electromechanical relay 25
embedded systems 10
Ethernet 39, 45, 56–58

FMS 7, 8, 102, 103, 104, 105, 107, 108, 109

hardware 15, 19, 21, 28, 38–58, 69, 106, 113, 117, 133
HVAC 83
hydraulic jacks 34–37, 50
hydraulic systems 33, 119

I2C 44, 63, 68
ICT/IT/information 9, 11
industrial robots 10, 48, 50, 121, 122, 130, 131
Intel Galileo 39, 40
intelligence 4, 5, 11, 14, 137
intelligent buildings 5, 9
Internet of Things 14
interoperability 14, 57
IoT 14

knowledge 12, 15, 137, 138, 139, 141, 148, 149, 150, 153

launching girder 34, 35, 36, 37
lead screw 32
light sensors 23, 59, 72–77, 82, 109
linear actuators 27, 31, 32, 49

machine learning 12, 49, 108, 115, 117–178, 199, 213–215, 244–248
mechanical actuators 24, 27
mechanization 10, 13, 113
microcontroller 24, 29, 38–40, 52, 54, 62, 63
multimeter 16, 17

open systems interconnection 55
optical encoders 28
OSI 56

parallel connection 19
PLC 38–39
potential difference 16
power source 16, 18, 19, 20, 21
prefabrication 9, 11, 120
Programmable Logic Controller 38–39
protocols 39, 44–47, 53–63, 68, 80
pulse width 29, 54, 55, 75
PWM 54

rack and pinion 33
Raspberry Pi 39, 41, 59
relay 25, 26, 27
router(s) 38, 39, 47, 48, 58, 59, 106
RPM 28
RS232 44, 45
RS485 47, 58

safety 6, 7, 9, 69, 119
SCI 44, 138
security 6, 13
sensors 14–24, 38–79, 90, 99
serial communication 24, 42–45, 60, 61
series connection 19, 20, 50
servomotor 28, 29, 32, 48, 49
smart building 4–9, 69
smart cities 9
solar 78, 79, 93, 94, 95, 96, 97, 98, 99
SPI 44, 61–63, 68
stepper motors 28, 30
switch 16, 25, 26

TCP/IP 39, 47, 56–59
thermostat 26

UART 45, 60
USB 21, 39, 40, 44–47, 56, 60

VOC 23
Volatile Organic Compounds 23
volt 17
voltmeter 16, 20

wireless 39, 47, 48, 50, 106, 121, 133

Zigbee 47, 106

9780367761103